KU-338-426

# Foundations of
# Mechanical Engineering

**JOIN US ON THE INTERNET VIA WWW, GOPHER, FTP OR EMAIL:**

WWW:      http://www.thomson.com
GOPHER: gopher.thomson.com
FTP:       ftp.thomson.com
EMAIL:    findit@kiosk.thomson.com

A service of **I(T)P**®

# Foundations of Mechanical Engineering

**Anthony Johnson**

and

**Keith Sherwin**

**CHAPMAN & HALL**

London · Weinheim · New York · Tokyo · Melbourne · Madras

**Published by Chapman & Hall, 2–6 Boundary Row, London SE1 8HN, UK**

Chapman & Hall, 2–6 Boundary Row, London SE1 8HN, UK

Chapman & Hall GmbH, Pappelallee 3, 69469 Weinheim, Germany

Chapman & Hall USA, 115 Fifth Avenue, New York, NY 10003, USA

Chapman & Hall Japan, ITP-Japan, Kyowa Building, 3F, 2-2-1 Hirakawacho, Chiyoda-ku, Tokyo 102, Japan

Chapman & Hall Australia, 102 Dodds Street, South Melbourne, Victoria 3205, Australia

Chapman & Hall India, R. Seshadri, 32 Second Main Road, CIT East, Madras 600 035, India

First edition 1996

© 1996 Anthony Johnson and Keith Sherwin

Typeset in $10\frac{1}{2}/12$ pt Times by Thomson Press (India) Ltd., New Delhi

Printed in Great Britain by the Alden Press, Osney Mead, Oxford

ISBN  0 412 61600 9

Apart from any fair dealing for the purposes of research or private study, or criticism or review, as permitted under the UK Copyright Designs and Patents Act, 1988, this publication may not be reproduced, stored, or transmitted, in any form or by any means, without the prior permission in writing of the publishers, or in the case of reprographic reproduction only in accordance with the terms of the licences issued by the Copyright Licensing Agency in the UK, or in accordance with the terms of licences issued by the appropriate Reproduction Rights Organization outside the UK. Enquiries concerning reproduction outside the terms stated here should be sent to the publishers at the London address printed on this page.
  The publisher makes no representation, express or implied, with regard to the accuracy of the information contained in this book and cannot accept any legal responsibility or liability for any errors or omissions that may be made.

A catalogue record for this book is available from the British Library

Library of Congress Catalog Card Number: 95-83526

∞ Printed on permanent acid-free text paper, manufactured in accordance with ANSI/NISO Z39.48-1992 and ANSI/NISO Z39.48-1984 (Permanence of Paper).

# Contents

# Preface

As the title implies, this book provides an introduction to the mechanical engineering disciplines of applied mechanics and thermofluid mechanics. From the list of contents it will be seen that the range of topics covered is that likely to be encountered by students on foundation and first-year courses in engineering.

Students are continually encouraged to 'self learn' and this book has been written with this in mind. The text is concise with a minimum of mathematical content. Each chapter defines a list of aims and concludes with a summary of the key equations introduced within the chapter. Outline solutions are provided for all the end of chapter problems so that students can check their own work.

In keeping with modern practice, 'velocity' has been used as a generic term to cover both speed and velocity. Strictly speaking, speed has magnitude and is a scalar quantity; velocity has both magnitude and direction and is a vector quantity. The use of velocity to cover both avoids confusion since the actual application will define whether it is a scalar or vector quantity.

There is a continuing debate as to the sign convention for the work output from a thermofluid system. We have retained the traditional convention of work done by a system being positive. This fits in with the intuitive concept of an engine as a device that produces positive work.

Many people have helped with this book and will recognize their own contribution. In particular, we wish to thank the Ford Motor Company Ltd for their permission to reproduce the drawing in Figure 1.1, Bethany Jenkins for doing such an excellent job in preparing the rest of the figures and diagrams and the publisher for allowing us to use data on the properties of metals and steam given in the appendices.

It is possible that some mistakes will have slipped through the proofreading stage. We would welcome feedback on these and the book as a whole.

Anthony Johnson
Keith Sherwin

# List of symbols

| | |
|---|---|
| $a$ | acceleration |
| $A$ | area, amplitude |
| $C$ | specific heat, couple |
| $C_d$ | coefficient of discharge |
| $C_P$ | specific heat at constant pressure |
| $C_V$ | specific heat at constant volume |
| $d$ | diameter |
| $E$ | modulus of elasticity |
| $f$ | skin friction coefficient |
| $F$ | force |
| $F_r$ | friction force |
| $g$ | gravitational acceleration |
| $h$ | enthalpy |
| $h_c$ | heat transfer coefficient |
| $I$ | moment of inertia, second moment of area |
| $J$ | second polar moment of area |
| $k$ | radius of gyration, thermal conductivity, spring stiffness |
| $l$ | lead of screw thread |
| $L$ | length |
| $m$ | mass |
| $\dot{m}$ | mass flow rate |
| $M$ | moment, molecular weight |
| $n$ | number of revolutions, factor of safety |
| $N$ | normal force |
| $P$ | pressure |
| $q$ | heat |
| $Q$ | rate of heat transfer |
| $r$ | radius |
| $R$ | gas constant |
| $R_0$ | universal gas constant |
| $s$ | distance |
| $t$ | time, thickness |
| $T$ | torque, temperature |
| $T_f$ | fluid temperature |
| $u$ | internal energy |
| $U$ | overall heat transfer coefficient |
| $v$ | velocity |
| $V$ | volume |

| | |
|---|---|
| $w$ | work done |
| $W$ | power |
| $x$ | dryness fraction |
| $x, y$ | distance in particular directions |
| $z$ | height |
| | |
| $\alpha$ | angular acceleration |
| $\gamma$ | index of adiabatic gas process $= C_P/C_V$ |
| $\delta$ | angle of friction |
| $\varepsilon$ | strain |
| $\theta$ | angular displacement |
| $\mu$ | coefficient of friction, viscosity |
| $v$ | Poisson's ratio |
| $\rho$ | density |
| $\sigma$ | direct stress |
| $\tau$ | shear stress |
| $\omega$ | angular velocity |

# Problem solving and basics $\boxed{1}$

## 1.1 AIMS

- To introduce mechanical engineering.
- To introduce the idea of using visual and mathematical models as a basis for problem solving.
- To describe the use of free-body diagrams and thermofluid systems as visual models of real situations.
- To discuss Newton's laws as a basis for mathematical modelling.
- To introduce the SI system of units.

## 1.2 MECHANICAL ENGINEERING

In order to study the foundations of mechanical engineering, it is first necessary to understand what is meant by the expression 'mechanical engineering'.

As a starting point, mechanical engineering can be defined as that branch of engineering concerned with the design, manufacture, installation and operation of engines and machines. One of the key functions of mechanical engineering is to design new products. To do so, the engineer must be able to analyse the product in order to assess its performance and to ensure that it is strong enough to perform its duty. As such, the analysis of mechanical engineering systems and components involves an understanding of the principles of the following topics in mechanical science:

- dynamics
- statics
- solid mechanics
- material science
- thermofluid mechanics.

These topics are introduced and discussed in the following chapters.

The term 'mechanical engineering' originated in the nineteenth century. Before that, there were two recognized branches of engineering, namely military engineering and non-military, i.e. civil, engineering. These were concerned with the building of roads, bridges and canals. The civil engineering activities were recognized by the formation in the UK of the Institution of Civil Engineers, the first professional engineering society, in 1818, with Thomas Telford as its first president.

With the spread of railways in the 1830s and 1840s, it was recognized that two different branches of engineering were involved. The building of bridges, viaducts and tunnels obviously involved civil engineering but the design and maintenance of steam locomotives involved a different set of skills. As a consequence, the Institution of Mechanical Engineers was established in Birmingham in 1847, with George Stephenson as its first president. The headquarters of the Institution were transferred from Birmingham to London in 1877.

Steam engines played a major role in the development of mechanical engineering throughout the nineteenth century, with application not only for transportation but also in industry. Even now, steam turbines are used for the generation of electricity in fossil fuelled and nuclear power stations.

During the twentieth century mechanical engineering has become much broader, embracing internal combustion engines, drive mechanisms, machine tools, air-conditioning plant and refrigeration systems. Modern society would be impossible without the products of mechanical engineering. In the home, a flick of a switch brings into operation devices that depend on electricity generated by means of a mechanical system in a thermal power station. Even the use of renewable energy sources using hydroelectric systems and wind turbines depends on mechanical engineering devices. Outside the home, all forms of transportation rely on engines, structures and drive systems created by mechanical engineers.

The above description gives a brief outline of mechanical engineering in order to set the scene for the topics discussed in the following chapters. If it gives the impression that mechanical engineering is a broad-based activity without which modern society could not functions, then it has achieved its purpose. Additional information on the history and development of mechanical engineering can be gained from Derry and Williams (1970) and Rolt (1960, 1967).

## 1.3  PROBLEM SOLVING

The whole purpose of studying mechanical engineering is to be able to solve problems. To do this, the student must have a good understanding of the physical behaviour of the device or situation under consideration. In order to be able to analyse such devices it is first necessary to be able to model the situation.

A **model** is a means of representing the real device or situation. The most familiar type of engineering model is probably the 'scale' model. For example, an aircraft manufacturer would not go to the expense of building a full-sized prototype without first carrying out tests on scale models in the wind tunnel to evaluate the performance characteristics.

Although scale models are used in mechanical engineering, their use is not as widespread as might be imagined. When designing a new internal combustion engine, for example, little of value can be gained from a test on a small-scale model of the engine. This is due to what is called the 'scale effect'. The friction and heat losses for a model are proportionally greater than for a full-sized engine. This is because the surface area of a component, as a ratio of its contained volume, increases as the component becomes smaller, resulting in proportionally greater losses. From the alternative point of view, losses are reduced as the size is increased. This is why some of the larger animals, such as elephants, have low thermal losses and tend to 'overheat'.

Scale models do not form part of the topics within this text; nevertheless, modelling is an essential part of problem solving and forms the basis for the subsequent discussion. The types of model that are used within this book to define and analyse mechanical engineering situations can be broadly classified as:

1. *visual model* – used to visualize the device or situation;
2. *mathematical model* – used to formulate equations in order to define the behaviour under working conditions.

Students tend to get worried by the term 'mathematical model' because it gives an impression of abstruse and complex formulae. Nothing could be further from the truth. While it is necessary to use formulae in order to analyse some mechanical engineering situations, the most complex situations are now solved using packages on the computer. The formulae found in the following chapters represent the essential basics; in other words, the foundations of mechanical engineering. While some mathematical knowledge is assumed on the part of the student, the use of calculus is avoided in the derivation of the formulae.

Using visual and mathematical models it is possible to outline an ordered approach to the solving of mechanical engineering problems. Although not all problems are the same, or are solved in the same way, it is necessary to have some idea of how to tackle problems. The following step-by-step approach provides a basis for solving problems.

1. *Understand the problem* – an obvious statement but one that needs emphasizing. Understanding means having a clear mental picture of what the problem entails and this means visualizing the physical situation involved.
2. *Draw a sketch or diagram* – in order to solve a problem it is useful to draw a visual model of the situation. This might simply be a line sketch or, alternatively, a more complex technical diagram, but whatever form the visual model takes it is necessary to concentrate the mind on the essential features of the situation.
3. *Create a mathematical model* – apply the basic laws to the situation defined in the visual model in order to create the equations or formulae necessary for solving the problem.
4. *Make assumptions* – where possible simplify the analysis to make it more applicable to a particular problem by ignoring any terms that have a negligible effect. For example, if the velocity of a particle is very low, its kinetic energy can be considered as insignificant.
5. *Analyse the problem* – use the simplified formula in a rearranged form to determine the unknown quantities within the problem.

6. *Check the answer* – check that the units are correct and that the numerical solution looks to be of the right magnitude.

Problem solving becomes easier to appreciate through the solution of actual problems, an illustration of the old saying that practice makes perfect.

The step-by-step approach to problem solving outlined above serves two functions. Firstly, it shows that problem solving is not a hit-or-miss activity but is achieved in an ordered manner. Secondly, steps 1–3 show that modelling forms the basis of problem solving. It therefore serves as an introduction to the following discussion on the particular types of visual model used in mechanical engineering problems.

## 1.4   DRAWING A DIAGRAM

Figure 1.1 shows a diagram of a four-cylinder spark ignition petrol engine. As such, the diagram shows a wealth of interesting technical detail. For example, it shows that the engine operates on the four-stroke principle. By studying the diagram closely other details can be seen that help to provide a more complete understanding of the whole system.

Now, the term 'system' is one that is used regularly in mechanical engineering, so this is a good point to explain what is meant by 'system' and, as a consequence, what is meant by 'component'.

Figure 1.1 represents an engineering system because it combines a large number of individual parts that interact in order to form a collective device. Therefore, a system is a combination of parts working together to achieve the desired outcome. In the case of an engineering system, such as the engine shown in Figure 1.1, each individual part, whether a piston or a single bolt, is a 'component' and goes to make up the whole.

The question naturally arises, if the engine shown in Figure 1.1 is a system, what is the car that the engine is installed in? The answer is that the car is also termed a system, although a rather more complex system than the engine. If the car is thought of as a system, then the engine can be considered to be a subsystem of the car. Taking, say, a piston–pin–connecting rod from the engine, these three separate components form a subsystem of the engine when joined together. Therefore, an engineering system can comprise any number of components, providing that the components are acting together.

Having now briefly defined an engineering system, it is possible to return to Figure 1.1 and consider whether this diagram can be used as a visual model in order to analyse the behaviour of the engine. The answer is no! Figure 1.1 contains a lot of visual detail but, as a basis for analysing the engine, it has two disadvantages:

1. It lacks numerical information. For example, there is no indication of the size of the engine.
2. It is too complex.

In order to visualize an engineering system or component it is necessary to make the diagram as simple as possible by isolating the situation under

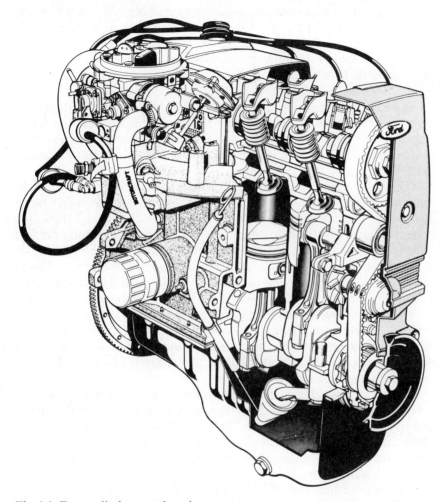

**Fig. 1.1** Four-cylinder petrol engine.

consideration from its surroundings. This can be done by using two alternative forms of visual model:

1. a free-body diagram
2. a thermofluid system.

These will be discussed in greater detail later in the text but, for the present, the following subsections give a brief introduction to each of these visual models.

### 1.4.1 Free-body diagram

By taking one component, say a connecting rod, from the engine shown in Figure 1.1 the behaviour and characteristics of that connecting rod can be analysed by isolating it from the rest of the engine. This means that the rod is assumed to be operating on its own.

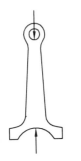

**Fig. 1.2** Free-body diagram of a connecting rod.

Clearly, this is impossible because the connecting rod has a function only through connecting the movement of the piston to the crankshaft. Nevertheless, the connecting rod can be considered on its own providing that the influence of the other components on the connecting rod is taken into account. This can be done by drawing a free-body diagram of the connecting rod, as shown in Figure 1.2.

By drawing a free-body diagram it is possible to concentrate on the essential features of the problem. In this case it means that the forces applied externally on the connecting rod, from the piston and crankshaft, can be drawn and this allows, say, the analysis of the stresses within the rod.

The forces shown in Figure 1.2 are somewhat simplistic and would, in practice, be complicated by the inertia of the rod and the forces being distributed over the bearing surfaces. Nevertheless, Figure 1.2 shows the salient features of a free-body diagram, which represents a body isolated from its surroundings with the appropriate applied external forces. To go into greater detail at this stage would mean pre-empting discussion given later in the text. However, it is necessary to clarify what is meant by 'body' in a free-body diagram.

A free-body diagram can be applied to a single component or to several components acting together. In the case of the former the boundary of the free-body is represented by the external shape of the component. In the case of the latter, the boundary of the free-body is represented by the outline shape of the system.

Figure 1.3 shows the free-body diagram for a piston and connecting rod assembly. Assuming that the piston is moving downwards due to gas pressure in the cylinder, the forces represented on the top of the piston are those due to the pressure. There will be a sideways force due to the reaction of the cylinder on the piston and, since there must be some friction due to the movement of the piston, a friction force.

What is particularly noticeable about the free-body diagram shown in Figure 1.3 is that no account is taken of internal forces within the system. Only the external forces are considered.

### Example 1.1

Draw the free-body diagram for a person on a simple plank bridge supported at either end.

Free-body diagram – the person will apply an external force on the plank due to his or her weight. This will be a downwards force. The two supports will apply upwards force to the plank and the resulting free-body diagram is shown in Figure 1.4.

**Fig. 1.3** Forces on a piston and connecting rod assembly.

### 1.4.2 Thermofluid system

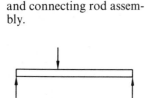

**Fig. 1.4**

Considering the engine shown in Figure 1.1, the space contained within a cylinder and piston assembly is occupied by a gas. Assuming that there

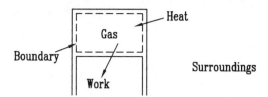

**Fig. 1.5** A thermofluid system.

is no flow of gas into, or out of, the cylinder, the gas is either doing work on the piston or having work done on it by the piston. To analyse the performance of the engine it is necessary to isolate the gas from the surrounding engine and simply consider the behaviour of the gas. This can be done by considering the contained gas as a thermofluid system.

Figure 1.5 shows the gas contained in the cylinder and piston assembly as a thermofluid system. In order to explain the term 'thermofluid system' it is necessary to recap on the previous discussions on a system. It was stated that a system consists of several interacting parts, or components. In the case of the gas considered in Figure 1.5, it is itself a system because it consists of many interactive parts, the parts being the molecules of which the gas is composed.

Thermofluid systems are discussed in more detail in Chapter 11, but the basic features are shown in Figure 1.5. The gas is isolated from the 'surroundings', in this case the rest of the engine, by a 'boundary' indicated by the dotted line. Across the boundary, energy can move in the form of 'heat' and 'work'. These energy transfers are analogous to the external forces applied to a free-body diagram; they are the link between the gas and the surroundings. If the gas is at a higher pressure than the surroundings and, as a result, causes the piston to move, then it is doing work. On the other hand, if the piston is moving to compress the gas, then there is work being done on the system.

If the gas is at a higher temperature than the surroundings, there is a transfer of heat from the gas. This is why, in an engine, the cylinder is surrounded by cooling water to provide a means of transferring heat away from the hot gases in the cylinder.

It should be emphasized that, in the study of thermofluid mechanics, 'heat' has a specific meaning. Heat is the term used to describe only that form of thermal energy that crosses the boundary.

A thermofluid system is defined by the extent of the boundary. For example, it is possible to draw a boundary around the whole of the engine shown in Figure 1.1 and define a thermofluid system as shown in Figure 1.6.

However, in the case of a complete engine, the system is slightly different from that shown in Figure 1.5 as there is movement of gas across the boundary. Air enters the engine from the surrounding atmosphere to both support combustion and act as the working fluid. After the engine has gone through two revolutions each cylinder has exhausted its spent

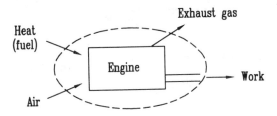

**Fig. 1.6** An engine as a thermofluid system.

gas to the atmosphere again. When there is a flow of any fluid across the boundary, the thermofluid system is termed an 'open' system. When there is no flow of fluid across the boundary, the thermofluid system is termed a 'closed' system.

It will be noticed that, in the case of the engine shown in Figure 1.6, there is a transfer of heat and work across the boundary. For the engine, the heat is transferred in the form of fuel entering the engine from the surroundings and being burnt internally. The work leaves the engine through the end of the crankshaft.

### Example 1.2

Define the type of thermofluid system that can be used to describe a potato being cooked in an open saucepan.

Thermofluid system – the system depends on where the boundary is drawn, as illustrated in Figure 1.7.

*Solution*

(a) Boundary A contains the potato and, since there is no flow of fluid across the boundary, the system is a closed system with heat transfer from the surrounding water to the potato.

(b) Boundary B contains both the potato and saucepan. If the water is not boiling, then the system is again a closed system with heat transfer from a gas, electric or solid fuel source. However, if the water is boiling, the vapour will cross the boundary and the system will then be an open system.

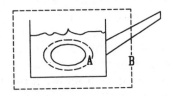

**Fig. 1.7**

## 1.5 NEWTON'S LAWS

Having briefly introduced visual modelling using free-body diagrams and thermofluid systems, it is necessary to lay the foundations for mathematical modelling of situations by considering the basic laws of Isaac Newton.

Generally, when Newton's laws are mentioned, they are taken to mean the three laws of motion. However, this ignores two other laws that are widely used in the analysis of mechanical engineering situations, namely:

1. law of gravitation
2. law of cooling.

### 1.5.1 Newton's laws of motion

As applying to a particle the three laws of motion can be expressed as:

1. A particle remains at rest or continues to move in a straight line with uniform velocity unless acted upon by an external force.
2. The acceleration of a particle is proportional to the external force acting on the particle and is in the direction of the force.
3. For every force acting on a particle there is an equal and opposite reactive force.

These laws are discussed in more detail in Chapter 3, so the present section simply gives a brief outline of what the laws mean.

The first law is the basis of the principle of **equilibrium** for a particle, irrespective of whether it is in a solid or fluid state. Equilibrium means that the particle is in a stable condition and that all the forces acting on the particle cancel out each other so that no resultant force remains. If there is a resultant force, the particle is no longer in equilibrium and the force causes the particle to accelerate, as stated in the second law.

When a particle, or body, has an external force applied to it, the result is that the particle accelerates in the direction of the applied force. The resulting acceleration is proportional to the magnitude of the external force and is governed by the relationship

$$F = m \times a \tag{1.1}$$

where  $F$ = external force,
$m$ = mass of the particle,
$a$ = resulting acceleration.

In more general terms this means that the product of the mass and the acceleration represents the **rate of change of momentum** of the particle. Since momentum is the product of the mass and the velocity, Newton's second law can be re-expressed in the form

$$F = \frac{\Delta(mv)}{\Delta t} \tag{1.2}$$

where $\Delta(mv)$ = incremental change in momentum,
$\Delta t$ = incremental change in time.

Equation (1.2) is particularly useful when considering the force resulting from the flow of a fluid as in, say, the thrust developed by a jet engine. In order to apply equations (1.1) and (1.2) it is necessary to have a consistent system of units with which to define the different variables. The SI system of units is discussed later in this chapter.

The third law states that if there is a force acting on a solid particle, or object, the object provides an equal force in the opposite direction to the applied force. Take, for example, a person standing. The person can stand in a stationary manner if the ground supports his or her weight. From the third law the ground provides an upwards force equal and opposite to the weight of the person. If the force provided by the ground were less than the weight, the person would sink as would be the case of a quicksand. If the force created by the ground were greater than the weight, then the person would move upwards and become airborne.

**Example 1.3**

If a rocket is moving in outer space, what happens when:

(a) the motor is working;
(b) the motor is switched off?

*Solution*

(a) The rocket motor creates a propulsive force and, from Newton's second law, the rocket will continue to accelerate.
(b) With the motor switched off, there is no external force applied and, from Newton's first law, the rocket will continue in a straight line at the same uniform velocity achieved at the end of the motor run.

### 1.5.2 Newton's law of gravitation

Newton defined his universal law of gravitation in 1687. Formally, it can be stated as:

> Any two bodies will attract each other with a force that is proportional to the product of their masses and inversely proportional to the square of the distance between their centres.

In practice, this means that the attractive force increases as the combined mass of the two bodies increases and reduces the further the two bodies are apart.

This is true irrespective of the size of the bodies, so that an object on the surface of the earth experiences a force due to the attraction of the earth. This force is termed the 'weight' of the object and, from the second law of motion, can be expressed as

$$F = m \times g \tag{1.3}$$

where  $F$ = the force due to gravitational attraction of the earth, the
        weight of the object,
      $m$ = mass of the object,
      $g$ = gravitational acceleration.

The value of $g$ varies slightly over the surface of the earth because of
the variation in the radius. Generally, at sea-level the value is taken as

$$g = 9.81 \text{ m/s}^2$$

Similarly, the earth will experience a force equal to the weight of the
object, but since it has negligible influence on the earth it can be ignored.
This is not the case where the masses of the two bodies are of the same
order of magnitude. The moon causes an attractive force on the earth
which is more noticeable because it results in the tidal movement of the
sea.

The mass of a body is an intrinsic characteristic of the body resulting
from its size and the density of the material contained within the body. Its
weight is dependent on the gravitational acceleration and varies with the
distance from the surface of the earth, but the mass remains the same.
When astronauts experience 'weightlessness' it does not mean that they
are outside the influence of the earth's gravitational attraction. Weightlessness, in this situation, is due to the forces on the astronaut, or spacecraft, being in equilibrium. The force due to gravity is balanced by a force
on the astronaut and the spacecraft, as a result of their velocity around
the earth. If a body is rotated at a velocity, the force on the body that
results from the rotation is called the 'centrifugal force'. It is analogous to
a conker being rotated at the end of a piece of string. The string provides
an inward force on the conker and the rotation provides an outward
centrifugal force that keeps the conker in equilibrium.

The gravitational attraction on a body is illustrated in Figure 1.8. The
diagram in Figure 1.8(a) shows what happens in reality. All the particles
that go to make up the body experience an attraction when operating in
a gravitational field. However, it is inconvenient to analyse a situation of
this nature because the influence of all the separate particles has to be
taken into account. Fortunately, the whole body can be considered as
a total entity by using a free-body diagram, as illustrated in Figure 1.8(b).
The total mass of the body is assumed to act through a single point in the
body, $G$, called the 'mass centre'. Therefore, the weight of a body can be
drawn as a single force on the free-body diagram acting through the mass
centre.

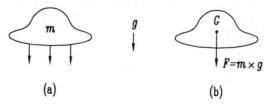

(a)                                         (b)

**Fig. 1.8** Gravitational attraction on a body.

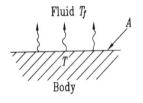

Fluid $T_f$

$T$

Body

**Fig. 1.9** Heat transfer from a body.

### 1.5.3 Newton's law of cooling

Newton formulated his law of cooling in 1701, rather later than his laws of motion or the law of gravitation. Basically, the law states that if a hot solid body is surrounded by a cooler fluid, the body will lose heat to the fluid. The rate at which heat is lost to the fluid is proportional to the difference in temperature between the body and the fluid. In fact, there can only be heat transfer as a result of a temperature difference. The heat transfer is also proportional to the surface area of the body.

The situation can be considered as shown in Figure 1.9. If the temperature of the body is $T$ and the fluid is at a lower temperature $T_f$, there will be heat transfer from the body to the fluid, which can be expressed as

$$Q = h_c A(T - T_f) \tag{1.4}$$

where  $Q$ = rate of heat transfer,
       $A$ = surface area of the body,
       $T - T_f$ = temperature difference between the body and the fluid.

The term $h_c$ is a measure of how rapidly the heat is transferred from the body and is determined by the characteristics of the fluid.

Equation (1.4) is the basic relationship for 'convective' heat transfer and is described in more detail in Chapter 15.

## 1.6  UNITS

In order to apply Newton's laws, or any other analytical relationships, it is necessary to have a coherent system of units to define the characteristics of the variables under consideration. Throughout this book the international system will be used. This is referred to as the Système International d'Unités, abbreviated to SI.

Within the SI system there are four basic units used in the analysis of mechanical systems; these being length, mass, time and temperature, defined in Table 1.1 with their SI units and symbols.

**Table 1.1** SI basic units

| Basic unit | Unit | Symbol |
|---|---|---|
| Length | metre | m |
| Mass | kilogram | kg |
| Time | second | s |
| Temperature | kelvin | K |

### 1.6.1  Derived units

The units of all variables used in mechanical engineering are derived from the four basic units.

Velocity is required to define the movement of a particle, whether the particle is a solid or a fluid. Velocity is the distance travelled during a given period of time:

$$\text{velocity} = \frac{\text{distance}}{\text{time}} = \frac{\text{length}}{\text{time}}$$

with units of

$$(\text{m})/(\text{s}) = \text{m/s}$$

Acceleration is the change of velocity with time:

$$\text{acceleration} = \frac{\text{velocity}}{\text{time}} = \frac{\text{length}}{(\text{time})^2}$$

with units of

$$(\text{m})/(\text{s}^2) = \text{m/s}^2$$

Volume defines the amount of space occupied by a quantity of fluid, or material, and has units of $(\text{m}^3)$. The density of the fluid, or material, contained within the space is defined as the mass of material contained within a unit volume:

$$\text{density} = \frac{\text{mass}}{\text{volume}}$$

with units of

$$(\text{kg})/(\text{m}^3) = \text{kg/m}^3$$

Force has units that are derived from Newton's second law of motion:

$$\text{force} = \text{mass} \times \text{acceleration}$$

with units of

$$(\text{kg}) \times (\text{m/s}^2) = \text{kg m/s}^2$$

This composite unit for force is called the newton, appropriately, and given the symbol N, where

$$N = \text{kg m/s}^2$$

The formal definition of the newton is the force required to accelerate a mass of 1 kg at a constant acceleration of $1 \text{ m/s}^2$ in the direction of the force.

Pressure, or stress, is defined as the force exerted on a unit area:

$$\text{pressure} = \frac{\text{force}}{\text{area}} = \frac{\text{force}}{(\text{length})^2}$$

with units of

$$(N)/(m^2) = N/m^2$$

This composite unit of pressure, or stress, is called the pascal and is given the symbol Pa, where

$$Pa = N/m^2$$

The formal definition of the pascal is the pressure, or stress, achieved when a force of 1 N acts uniformly over an area of 1 m$^2$.

Energy can be defined in the same units as work, since work is one form of energy. The other forms of energy include potential energy, kinetic energy and heat. From basic physics, work is given by

$$work = force \times distance \text{ in the direction of the force}$$
$$= force \times length$$

with units of

$$(N) \times (m) = N\,m$$

This composite unit of energy is called the joule and is given the symbol J, where

$$J = N\,m$$

The formal definition of the joule is the energy required when a force of 1 N acts through a distance of 1 m.

Power is the rate of doing work. In other words, it is the work done during a unit period of time:

$$power = work/time$$

with units of

$$(J)/(s) = J/s$$

The unit for power is called the watt and is given the symbol W, where

$$W = J/s$$

Since heat and work are both forms of energy, the rate at which heat is transferred must have the same units as power and is also designated by the watt.

The most commonly used derived units are summarized in Table 1.2.

**Table 1.2** Derived units

| Variable | Unit | Symbol |
|---|---|---|
| Force | newton | N |
| Pressure, stress | pascal | Pa |
| Work, energy | joule | J |
| Power | watt | W |

## Example 1.4

Express the watt, the unit of power, in the SI base units.

*Solution*

$$W = \frac{J}{s} = \frac{N\,m}{s} = \left(kg\frac{m}{s^2}\right)\frac{m}{s}$$

Therefore

$$W = \frac{kg\,m^2}{s^3}$$

## Example 1.5

A rocket having a mass of 4000 kg is propelled in outer space by a motor that burns 12 kg/s of fuel and oxygen, and achieves an exhaust velocity of 1500 m/s. Ignoring the change in mass of the rocket as the fuel in burnt, what is the acceleration of the rocket?
  Visual model – given in Figure 1.10.

*Solution*

The magnitude of the thrust of the motor can be found from equation (1.2):

$$F = \frac{\Delta(mv)}{\Delta t}$$

and for constant exhaust velocity this can be re-expressed as

$$F = \frac{\Delta m}{\Delta t} \times v$$

$$12 \times 1500 = 18\,000\,N$$

The acceleration of the rocket can be found from equation (1.1):

$$F = m \times a$$

and

$$a = \frac{F}{m} = \frac{18\,000}{4000} = 4.5\,m/s^2$$

**Fig. 1.10**

### 1.6.2 Temperature

Since temperature is one of the basic units it is important to understand what it means. Unfortunately, it is not easy to give a precise definition other than temperature is a measure of the 'hotness' or 'coldness' of a body or fluid.

Hotness or coldness are very subjective concepts so it is necessary to have a common temperature scale for measurement. The SI temperature scale is the Celsius scale. This is defined by means of two fixed temperatures, the freezing point and boiling point of water under standard atmospheric conditions. These are designated by $0\,°C$ and $100\,°C$.

However, the unit of temperature given in Table 1.1 is the kelvin, K, based on the **absolute temperature** scale. The value of K is related to the Celsius scale in that an increase in temperature of 1 K is equal to an increase of $1\,°C$. The difference between the two is that each scale is related to a different value of zero.

The lowest possible temperature in the universe is **absolute zero** at a temperature of $-273.15\,°C$. This is used as the datum for the absolute scale so that

$$0(K) = -273.15(°C)$$

In the analysis of mechanical engineering situations it is sufficiently accurate to take absolute zero to the nearest whole number, i.e. $-273\,°C$, and this value can be used to convert temperatures on the Celsius scale to absolute kelvin:

$$T(K) = T(°C) + 273 \tag{1.5}$$

It follows that a temperature of, say, $25\,°C$ can be expressed as either $25\,°C$ or as 298 K. In practice, temperatures are generally expressed in $°C$ because this is the value that is actually read from a temperature-measuring device such as a thermometer. For some analyses it is then necessary to convert from temperatures in $°C$ to kelvin.

Since a particular temperature can be expressed in either $°C$ or K, it raises the question of which temperature unit to use for a temperature difference. For example, Newton's law of cooling as expressed in equation (1.4) relies on the temperature difference between a surface and a fluid. Suppose the surface is at $80\,°C$ and the fluid is at $20\,°C$, the difference between them is 60, but 60 what? The convention is to quote temperature differences in K and the reasoning behind this becomes apparent when the two original temperatures are converted to absolute temperatures:

$$\text{temperature difference} = (80 + 273) - (20 + 273)$$
$$= 60\,K$$

### Example 1.6

A domestic radiator has a height of 600 mm and a horizontal length of 1 m. If the water in the radiator is at $70\,°C$ and the surrounding air is at

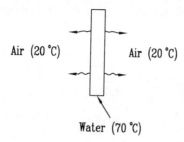

Air (20 °C)          Air (20 °C)

Water (70 °C)

**Fig. 1.11**

20 °C, find the rate of heat transfer from the surface. Take the value of $h_c$ as 6 W/m² K.

Visual model – taking a cross-section through the radiator, there will be heat transfer from both sides (Figure 1.11).

*Solution*

The rate of heat transfer can be found from equation (1.4):

$$Q = h_c A(T - T_f)$$

Checking on units:

$$Q = \frac{W}{m^2\,K}\,m^2\,K = W$$

The total surface area of the radiator is

$$A = 2 \times (0.6 \times 1) = 1.2\,m^2$$

Substituting in equation (1.4):

$$Q = 6 \times 1.2 \times (70 - 20)$$
$$= 360\,W$$

### 1.6.3 Unit prefixes

Some of the SI units represent quite small quantities. For example, a pressure of 1 Pa is equivalent to the effect of a layer of water just 0.1 mm thick and any workable pressure involves the use of large numbers. Atmospheric pressure is equal to a value in the region of 101 300 Pa. Such large numbers are inconvenient when carrying out an analysis and it is more sensible to use multiples of the unit to make the numerical calculations more manageable. Table 1.3 gives a list of the multiplying factors in common use with SI.

From the pressure quoted above, the value can be expressed in kilopascals (kPa) where

$$101\,300\,Pa = 101.3\,kPa$$

**Table 1.3** Multiples in the SI system

| Multiple | Prefix | Symbol |
|----------|--------|--------|
| $10^9$ | giga | G |
| $10^6$ | mega | M |
| $10^3$ | kilo | k |
| $10^2$ | centi | c |
| $10^{-3}$ | milli | m |
| $10^{-6}$ | micro | $\mu$ |

The prefixes given in Table 1.3 are applied to the derived units given in Table 1.2, but are not applied to all units.

Prefixes are not applied to the units of temperature, so temperatures are quoted in °C or K irrespective of the value.

Time does not fit into the system of prefixes. Although the second (s) is the basic unit of time in the SI system, long periods are still expressed in terms of the minute, hour or day:

$$1\,\text{min} = 60\,\text{s}$$
$$1\,\text{h} = 60\,\text{min} = 3600\,\text{s}$$
$$1\,\text{day} = 24\,\text{h} = 86\,400\,\text{s}$$

For lengths, British Standards advocate the use of either the millimetre (mm) or metre (m) and engineering drawings are presented in the two units. However, when it comes to volumes, the difference between a cubic millimetre (mm$^3$) and a cubic metre (m$^3$) is too great for practical purposes. Therefore, alternative units of volume are used. The cubic capacity of a car engine is quoted in cubic centimetres (cm$^3$ or cc). For liquids, volumes are sometimes expressed in litres, given the symbol l, where

$$1\,\text{l} = 10^{-3}\,\text{m}^3$$

## 1.7 SUMMARY

Key equations that have been introduced in this chapter are as follows.
Newton's second law of motion:

$$F = m \times a \tag{1.1}$$

Second law expressed as a rate of change of momentum:

$$F = \frac{\Delta(mv)}{\Delta t} \tag{1.2}$$

Newton's law of cooling:

$$Q = h_c A(T - T_f) \tag{1.4}$$

Conversion from Celsius temperatures to absolute temperatures:

$$T(K) = T(C^\circ) + 273 \qquad\qquad (1.5)$$

## 1.8   PROBLEMS

1. Sketch a free-body diagram of an aircraft cruising at an altitude of 7000 m.
2. Can a human being be represented as a thermofluid system?
3. Express the joule, the unit of energy, in terms of the basic units of length, mass and time.
4. Express a time period of 3 h 21 min in the basic unit of time, using a suitable prefix.
5. Find the power required to lift several people with a combined mass of 300 kg through a height of 50 m in 10 s. Assume $g = 9.81 \, \text{m/s}^2$.
6. The temperature of a human body is about 37 °C. Express this temperature as an absolute temperature.

# Motion of bodies $\boxed{2}$

## 2.1 AIMS

- To introduce the concepts of linear and angular motion.
- To explain the relationships between displacement, velocity and acceleration.
- To explain the relationship between absolute and relative velocities.
- To define the equations used to analyse linear and angular motion.
- To introduce an approach by which linear and angular motion problems can be analysed.
- To explain related topics such as 'falling bodies', 'trajectories' and vector methods.

## 2.2 INTRODUCTION TO MOTION

When traffic lights turn to green a car will move away with increasing velocity. The car will cover a distance in a particular direction and will possess a particular velocity at any instant. During this process the car possesses the three basic constituents of motion, namely: displacement, velocity and acceleration. It should be noted that since the car runs on wheels, these will also be in motion and therefore possess displacement, velocity and acceleration. However, the car moves in a linear direction, while the wheels move in an angular direction.

### 2.2.1 Displacement

If a man walks 10 km, there is an indication of the distance between the start position and the final position, but there is no indication of the direction. The 10 km is merely the **distance** covered and, as such, is a scalar quantity, i.e. possessing magnitude only. Displacement, however, implies a change in position or movement over a distance and gives the position and direction from the start point. Thus **displacement** is a vector quantity possessing both magnitude and direction.

Figure 2.1 gives an example of a man who walks 3 km east then 4 km north. He has actually walked a distance of 7 km but has been displaced from his start point by only 5 km.

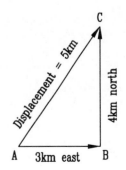

**Fig. 2.1** Displacement diagram.

### 2.2.2 Velocity

Velocity is the value of displacement measured over a period of time. It is the rate over which a distance/displacement is traversed. The magnitude of velocity is often expressed in convenient units such as kilometres per hour or miles per hour; however, these should be regarded as observation and comparison units. For analysis purposes velocity is better expressed in SI units of m/s.

### 2.2.3 Average velocity

Consider a car travelling between two towns at an average velocity of 50 km/h. On the journey the car will have stopped at traffic lights, crawled in traffic queues and 'speeded up' on fast stretches of road. It would be difficult to record the variations in velocity throughout the journey but average velocity can be considered as follows:

$$\text{average velocity} = \frac{\text{total distance}}{\text{total time taken}}$$

say

$$\frac{75\,\text{km}}{1.5\,\text{h}} = 50\,\text{km/h}$$

The average velocity ignores the variations in the actual velocity and gives a value which assumes the whole journey to have been undertaken at a constant velocity of 50 km/h.

An example of an average velocity calculation is when 'lap time' is recorded for a racing car completing laps on a racing circuit. The lap time is taken from the start of the lap to the completion of the lap and can then be used to calculate the average velocity using the known distance around the circuit.

**Constant velocity** is a special value because it assumes that a body moves over equal distances in equal intervals of time. In terms of the car considered above, it would need to start instantly, move in a given direction at 50 km/h and continue at that velocity, without variation, until it reached its destination, where it would instantly stop.

This situation is obviously not practical but for analysis it is sometimes useful to consider. Perhaps the closest physical situations are those where a near constant velocity has been reached and the time is measured over a particular distance. Typical examples include water or land speed record attempts where the distance is usually 'the measured mile'.

Average velocity and constant velocity are measured in the same way:

$$v = \frac{s}{t} = \frac{\text{displacement}}{\text{time}} \tag{2.1}$$

where $v$ = average/constant velocity (m/s),
      $s$ = displacement (m),
      $t$ = time (s).

### 2.2.4 Displacement–time graphs

Velocity can be represented in terms of a simple graph as shown in Figure 2.2. This shows the velocity of a body travelling 4.0 m in 80 s at a constant velocity. The velocity is given by the slope of the line and can be considered as follows:

$$v = \frac{\text{displacement}}{\text{time}} = \frac{s}{t} = \frac{4.0}{80} = 0.05 \, \text{m/s}$$

### Example 2.1

An aircraft travels a distance of 9000 km at a constant velocity in 12 h. Find:

(a) the aircraft velocity in m/s;
(b) the distance travelled in 30 min;
(c) the time taken to travel 120 km.

*Solution*

The problem can be solved using equation (2.1) as follows:

(a) $v = \dfrac{s}{t}$

Now $12 \, \text{h} = 12 \times 3600 = 43\,200 \, \text{s}$, so

$$v = \frac{9000 \times 1000}{43\,200} = 208.33 \, \text{m/s}$$

(b) $v = \dfrac{s}{t}$ so that $s = v \times t$

$30 \, \text{min} = 30 \times 60 = 1800 \, \text{s}$

$$s = 208.33 \times 1800 = 375\,000 \, \text{m} \quad \text{or} \quad 375 \, \text{km}$$

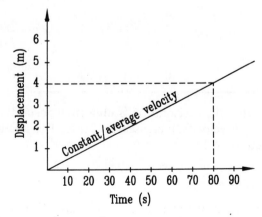

**Fig. 2.2** Displacement–time graph.

(c) $v = \dfrac{s}{t}$ so that $t = \dfrac{s}{v}$

$$t = \frac{120\,000}{208.33} = 576\,\text{s} \quad \text{or } 9\,\text{min } 36\,\text{s}$$

### 2.2.5 Acceleration

When a car sets off from traffic lights the driver depresses the 'accelerator' and the car steadily increases velocity. In slowing the car the driver depresses the brake pedal, which can be considered as a 'decelerator'. This example serves to define the terms acceleration and deceleration; however, other descriptive terms for deceleration are retardation and negative acceleration.

Acceleration is the change in velocity compared to advancing time, or

$$a = \frac{s}{t} \times \frac{1}{t} = \frac{s}{t^2}$$

where $a$ = acceleration in units of m/s$^2$.

### Example 2.2

The velocity of a car on a straight level road increases by 2.0 m/s every second as it accelerates from standstill until it reaches 40 m/s. Tabulate the velocity second by second.

*Solution*

The answer can be tabulated as follows:

| Time (s) | Velocity (m/s) | Acceleration (m/s) (Increase in velocity/s) |
|---|---|---|
| 0 | 0 | 2.0 |
| 1 | 2 | 2.0 |
| 2 | 4 | 2.0 |
| 3 | 6 | 2.0 |
| ↓ | ↓ | ↓ |
| 20 | 40 | 0 |

It can be seen that the acceleration is uniform (constant) and as every second ticks by the velocity increases by 2 m/s. The acceleration can, therefore, be seen to be 2 m/s$^2$.

### 2.2.6 Velocity–time graphs

Acceleration can be defined as the rate of increase of velocity. It can be represented on a velocity–time graph as shown in Figure 2.3.

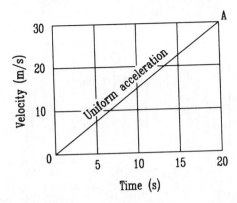

**Fig. 2.3** Velocity–time graph.

The velocity–time graph is a progression from the displacement–time graph shown in Figure 2.2. Instead of displacement, velocity is measured on the vertical axis. When velocity is plotted against time, the graph line represents acceleration. Figure 2.3 shows a straight graph line indicating uniform acceleration; however, in practice, acceleration can also vary.

It should be noted that the area under the graph line represents displacement and can be calculated by determining areas directly from the graph but also by considering equation (2.1):

$$v = \frac{s}{t}$$

or, transposing, $s = v \times t$. For the situation shown in Figure 2.3:

$$s = \text{area under graph} = \frac{30 \times 20}{2} = 300\,\text{m}$$

### 2.2.7 Mid-ordinate rule

If acceleration varies, the plot also varies, and it is not as straightforward to analyse as uniform acceleration. Sometimes the plot may be described by an equation, in which case the displacement, i.e. the area under the graph line, may be calculated using integral calculus. Often, however, it is more convenient to use a method of numerical integration such as the mid-ordinate rule.

Using the velocity–time graph shown in Figure 2.4, divide the base into several equal intervals. In this case the graph is divided into 10 intervals. The mid-ordinates are drawn vertically towards the graph line and measured.

The average velocity is found as follows:

$$\bar{v} = \frac{v_1 + v_2 + v_3 + v_4 + v_5 + v_6 + v_7 + v_8 + v_9 + v_{10}}{10}$$

Note that values of average velocity are defined as $\bar{v}$.

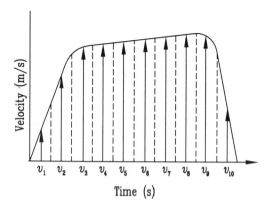

**Fig. 2.4** The mid-ordinate rule.

Once the average velocity has been determined, the displacement can be found by using the total time $t$ from the base line of the graph. From equation (2.1)

$$v = \frac{s}{t}$$

so that

$$s = vt = \text{area enclosed by graph} \times \text{time}$$

or

$$s = \bar{v}t = \text{average velocity} \times \text{time}$$

or

$$s = \bar{v} \times t$$

It should be noted that greater accuracy can be achieved by taking a greater number of ordinates.

## 2.3 EQUATIONS OF MOTION

Using the variables already defined it is possible to analyse linear motion using:

- displacement, $s$ (m)
- initial velocity, $v_1$ (m/s)
- final velocity, $v_2$ (m/s)
- acceleration, $a$ (m/s$^2$)
- time, $t$ (s).

A car travels along a road at a steady velocity and suddenly accelerates at a constant rate in order to overtake a slower car. The motion can be expressed graphically as shown in Figure 2.5.

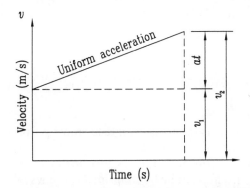

**Fig. 2.5** Uniform acceleration shown on a velocity–time graph.

Referring to Figure 2.5, the following observations can be made:

change of velocity $= v_2 - v_1$

This change of velocity can be related to the acceleration because

acceleration $= a =$ rate of change of velocity

$$= \frac{\text{change of velocity}}{\text{time taken}}$$

$$= \frac{v_2 - v_1}{t}$$

and transposing gives

$$v_2 = v_1 + at \tag{2.2}$$

This equation can be compared directly with the diagram in Figure 2.5.

Figure 2.5 shows that the velocity increases at a uniform rate between $v_1$ and $v_2$. The average velocity can therefore be described as

$$\bar{v} = \frac{(v_1 + v_2)}{2}$$

and if $s$ is the distance travelled during that period, equation (2.1) may be used in the form

$$s = vt$$
$$= \text{average velocity} \times \text{time}$$

In other words

$$s = \bar{v} \times t$$

and

$$s = \frac{(v_1 + v_2)t}{2} \tag{2.3}$$

From equation (2.2), it is possible to substitute for $v_2$ in equation (2.3):

$$s = \frac{(v_1 + v_1 + at)t}{2}$$

and rearranging gives

$$s = v_1 t + \tfrac{1}{2} a t^2 \qquad\qquad (2.4)$$

The above equations are adequate for most purposes but there often arise situations where time $t$ is the unknown. Since equations (2.2)–(2.4) all possess $t$ an expression is now needed for $v_2$ in terms of $v_1$, $a$ and $s$, which excludes $t$. This is done by squaring both sides of equation (2.2) and substituting for $s$ from equation (2.4). From equation (2.2)

$$v_2 = v_1 + at$$

and squaring gives

$$v_2^2 = (v_1 + at)^2$$

so that

$$v_2^2 = v_1^2 + 2v_1 at + a^2 t^2$$

or

$$v_2^2 = v_1^2 + 2a(v_1 t + \tfrac{1}{2} a t^2)$$

But $(v_1 t + \tfrac{1}{2} a t^2)$ is equal to $s$ from equation (2.4):

$$v_2^2 = v_1^2 + 2as \qquad\qquad (2.5)$$

The equations (2.2)–(2.5), derived above, form the basis of all motion studies at this level and can easily be manipulated to cope with a range of situations, including falling bodies and rotary motion, as explained later.

### Example 2.3

A train has a uniform acceleration of $0.2\,\mathrm{m/s^2}$ along a straight track. Find:

(a) the velocity after an interval of $16\,\mathrm{s}$ when the train starts from a standstill;
(b) the time required to attain a velocity of $50\,\mathrm{km/h}$;
(c) the distance travelled from standstill to when the train attains a velocity of $50\,\mathrm{km/h}$;
(d) the time taken for the velocity to increase from 30 to $50\,\mathrm{km/h}$;
(e) the distance travelled during the change in velocity from 30 to $50\,\mathrm{km/h}$.

*Solution*

(a) The velocity can be found using equation (2.2):

$$v_2 = v_1 + at$$

Now $a = 0.2\,\mathrm{m/s^2}$, $v_1 = 0$, so that

$$v_2 = 0 + (0.2 \times 16)$$
$$= 3.2\,\mathrm{m/s}$$

(b) The time taken can be found using equation (2.2):

$$v_2 = v_1 + at$$

Now $a = 0.2\,\mathrm{m/s^2}$, $v_1 = 0$ and

$$v_2 = 50\,\mathrm{km/h} \quad \text{or} \quad \frac{50\,000}{60 \times 60} = 13.9\,\mathrm{m/s}$$

and rearranging to isolate $t$, gives

$$t = \frac{v_2}{a} = \frac{13.9}{0.2} = 69.5\,\mathrm{s}$$

(c) The distance travelled in reaching $50\,\mathrm{km/h}$ can be found using equation (2.3):

$$s = \frac{(v_1 + v_2)t}{2}$$

Now $v_2 = 13.9\,\mathrm{m/s}$, $v_1 = 0$, $t = 69.4\,\mathrm{s}$, so that

$$s = \frac{(0 + 13.9)69.5}{2} = 483.03\,\mathrm{m}$$

(d) The time taken to change velocity from 30 to $50\,\mathrm{km/h}$ can be found using equation (2.2):

$$v_2 = v_1 + at$$

Now $v_2 = 13.9\,\mathrm{m/s}$, $a = 0.2\,\mathrm{m/s^2}$,

$$v_1 = 30\,\mathrm{km/h} \quad \text{or} \quad \frac{30\,000}{3600} = 8.33\,\mathrm{m/s}$$

and rearranging to isolate $t$ gives

$$t = \frac{v_2 - v_1}{a}$$

$$= \frac{13.9 - 8.33}{0.2} = 27.85\,\mathrm{s}$$

(e) The distance travelled during the change of velocity from 30 to $50\,\mathrm{km/h}$ can be found using equation (2.3):

$$s = \frac{(v_1 + v_2)t}{2}$$

Now $t = 27.85\,\mathrm{s}$, $v_1 = 8.33\,\mathrm{m/s}$, $v_2 = 13.9\,\mathrm{m/s}$, so that

$$s = \frac{(8.33 + 13.9)27.85}{2} = 309.6\,\mathrm{m}$$

### Example 2.4

A lift rises 200 m in 12 s. For the first quarter of the distance the lift is uniformly accelerated, and for the last quarter, uniformly retarded. For the rest of the run the velocity is constant. Find the maximum velocity.

   Velocity–time diagram – the distance travelled is equal to the area under the velocity–time graph (Figure 2.6).

*Solution*

The value of $v_{max}$ can be found by relating the distance travelled to the areas under the velocity–time graph:
a–b: using equation (2.3)

$$s = \frac{(v_1 + v_2)t}{2}, \quad \text{but} \quad v_1 = 0$$

$$S_{ab} = \frac{v_{max} \times t_{ab}}{2}$$

and

$$t_{ab} = \frac{2s_{ab}}{v_{max}} = \frac{100}{v_{max}} \tag{i}$$

b–c: using equation (2.1)

$$v = \frac{s}{t}$$

$$S_{bc} = v_{max} \times t_{bc}$$

and.

$$t_{bc} = \frac{S_{bc}}{v_{max}} = \frac{100}{v_{max}} \tag{ii}$$

c–d: using equation (2.3)

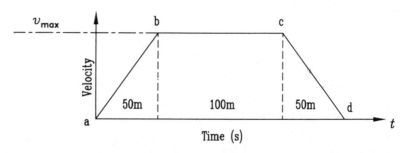

Fig. 2.6

$$s = \frac{(v_1 + v_2)t}{2}, \quad \text{but} \quad v_2 = 0$$

$$S_{cd} = \frac{v_{max} \times t_{cd}}{2}$$

and

$$t_{cd} = \frac{100}{v_{max}} \tag{iii}$$

Summing equations (i)–(iii) gives

$$t_{ab} + t_{bc} + t_{cd} = \frac{100 \times 3}{v_{max}}$$

Substituting for the total time taken

$$12 = \frac{3 \times 100}{v_{max}}$$

and

$$v_{max} = \frac{300}{12} = 25\,\text{m/s}$$

## Example 2.5

A motor cycle starts from rest at $s = 0$ and travels along a straight, level road at a velocity described in the $v$–$t$ graph (Figure 2.7). Determine the total distance travelled when the motor cycle stops at $t = 15\,\text{s}$.

*Solution*

The displacement can be found by using equation (2.3):

$$s = \frac{(v_1 + v_2)t}{2} \quad \text{for each area}$$

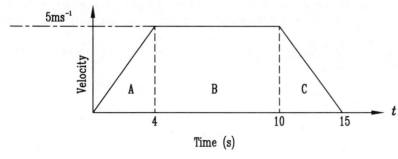

**Fig. 2.7** Velocity–time graph.

Area a:

$$s = \frac{(v_1 + v_2)t}{2} = \frac{v_2 t}{2} = \frac{5 \times 4}{2} = 10 \, \text{m}$$

Area b:

$$s = \frac{(v_1 + v_2)t}{2}$$

but $v_1 = v_2$ so that

$$s = \frac{2 \times v_1 t}{2} = 5(10 - 4) = 30 \, \text{m}$$

Area c:

$$s = \frac{(v_1 + v_2)t}{2} = \frac{v_1 t}{2} = \frac{5(15 - 10)}{2} = 12.5 \, \text{m}$$

*Note:* $v_1 = 5 \, \text{m/s}$

total displacement $= 10 + 30 + 12.5 = 52.5 \, \text{m}$

## 2.4 FALLING BODIES

A huge mass such as the earth attracts other masses such as cars, bricks, people, etc. and makes them stay on the earth's surface. If objects are allowed to fall, they will accelerate at a constant rate, termed the gravitational acceleration and designated by $g$. This is the constant acceleration due to gravitational attraction. The gravitational value $g$ varies at different points on the earth's surface but it is generally taken to be $9.81 \, \text{m/s}^2$.

If a feather and a steel ball are dropped from the same height at the same time, the steel ball will hit the ground first. Both objects will be attracted towards the ground at the same rate; however, air resistance on the feather prevents it from falling towards the earth with the same acceleration. If the same experiment were to be performed on the moon, where there is negligible atmosphere, both the ball and the feather would reach the surface at the same time.

In some instances air resistance may be a significant factor, as in the case of a parachutist, but in many cases the value is so small it may be ignored.

The motion of falling bodies can be analysed using equations (2.2), (2.3), (2.4) and (2.5).

### Example 2.6

A stone is dropped from the parapet of a bridge into the water below. The stone takes 4 s to fall from the moment of release to the moment when it hits the water. Estimate the height of the bridge.

*Solution*

The height of the stone upon release can be found using equation (2.4):

$$s = v_1 t + \tfrac{1}{2} a t^2$$

but $v_1 = 0$ and $a = g = 9.81 \, \text{m/s}^2$, so

$$s = \frac{9.81 \times 4^2}{2} = 78.5 \, \text{m}$$

**Example 2.7**

A stone is dropped from a 100 m high tower. Assuming $g = 9.81 \, \text{m/s}^2$ and air resistance to be negligible, find:

(a) the time taken to reach the ground;
(b) the velocity of the stone when it hits the ground;
(c) the distance through which the stone falls during the first 3 s.

*Solution*

(a) The time taken to reach the ground can be found using equation (2.4):

$$s = v_1 t + \tfrac{1}{2} a t^2$$

but $s = 100 \, \text{m}$, $a = g = 9.81 \, \text{m/s}^2$, $v = 0$, so transposing for $t$ and noting that $v_1 t = 0$ gives

$$t = \sqrt{\frac{2s}{g}} = \sqrt{\frac{2 \times 100}{9.81}} = 4.5 \, \text{s}$$

(b) The velocity, $v_2$, upon hitting the ground, can be found using equation (2.2):

$$v_2 = v_1 + at$$

but $v_1 = 0, t = 4.5 \, \text{s}$ and $a = g = 9.81 \, \text{m/s}^2$, so that

$$v_2 = 0 + (9.81 \times 4.5) = 44.15 \, \text{m/s}$$

(c) The distance the stone falls during the first 3 s can be found using equation (2.4):

$$s = v_1 t + \tfrac{1}{2} a t^2$$

but $v_1 = 0, t = 3 \, \text{s}$, $a = g = 9.81 \, \text{m/s}^2$, so that

$$s = 0 + \frac{9.81 \times 3^2}{2} = 44.15 \, \text{m}$$

## 2.5 RELATIVE VELOCITY

Velocity is said to be relative when the velocity of one object is observed from a defined viewpoint. This may be stationary or may possess its own

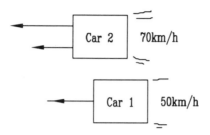

**Fig. 2.8** Relative velocity (direction).

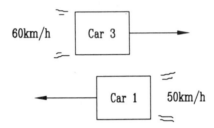

**Fig. 2.9** Relative velocity (opposing directions).

velocity, perhaps in a different direction from the velocity of the object. An example of this can be seen in Figure 2.8.

On a motorway, car 1 travels at 50 km/h and is overtaken by car 2 travelling at 70 km/h. The relative velocity is the velocity car 2 appears to be travelling when observed from car 1.

$$\text{relative velocity} = \text{velocity of car 2} - \text{velocity of car 1}$$
$$= 70 - 50 = 20 \text{ km/h}$$

i.e. car 2 is travelling 'relatively' 20 km/h faster than car 1.

Figure 2.9 shows car 1, again travelling at 50 km/h, in a positive direction. Car 3 is travelling at 60 km/h in the opposite direction (in a negative direction). The relative velocity is the velocity at which car 3 appears to be travelling when observed from car 1. The relative velocity in this case is the algebraic subtraction of the velocities of car 1 and car 3. This takes into account the negative direction of car 3, as shown below:

$$\text{relative velocity} = \text{velocity of car 1} - (-\text{ velocity of car 3})$$
$$= 50 - (-60) = 110 \text{ km/h}$$

### 2.5.1 Absolute velocity

Hitherto, it has been assumed that all velocities have been **absolute velocities**, that is, all measurements and observations were made as if the observer were stationary on the earth's surface while watching the moving body. This absolute velocity is a special case of relative velocity since the observations are made from the earth's surface, which is considered to be a fixed reference point.

### 2.5.2 Resultant of two velocities

Velocity possesses magnitude and direction and can, therefore, be represented in terms of a vector.

Two velocities which act at right angles to each other can be represented by a single vector, indicating the combined direction and velocity. Early aircraft navigation relied on this method. Before navigational aids were used to plot the position of aircraft the early flyers plotted the air speed as a vector on a chart and estimated the effects of side winds on the aircraft. These were drawn on the chart as vectors. The resultant velocity showed the true speed and direction of the aircraft.

### Example 2.8

An ocean linear steams 20 knots due north and is affected by 5 knot currents acting normal to the ship. (A knot is a nautical sea mile/h.) Determine the magnitude and direction of the resultant velocity using:

(a) a vector diagram
(b) trigonometric methods.

*Solution*

(a) Vector diagram – since velocity is a vector quantity the velocity of the ship and the current can be drawn to scale (Figure 2.10).
(b) The magnitude of the resultant velocity can be calculated using Pythagoras' theorem:

$$a^2 = b^2 + c^2$$
$$a = \sqrt{(20^2 + 5^2)} = 20.6 \, \text{knots}$$

The direction of the resultant can be found using trigonometry:

$$\tan \alpha = \frac{b}{c} = \frac{20}{5} = 76°, \text{ i.e. } 76° \text{ north of east}$$

*Note:* The above example is straightforward since the current and ship's velocity vectors lie at 90° to each other. In practice, however, vectors rarely lie in such convenient directions. In such cases vector methods may still be used, with other trigonometric methods such as the sine rule and the cosine rule.

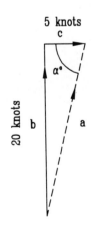

Fig. 2.10

### Example 2.9

An aircraft flies across the equator in a north-easterly direction at 30° to the equator. Its velocity is 400 km/h. The wind velocity, flowing in an

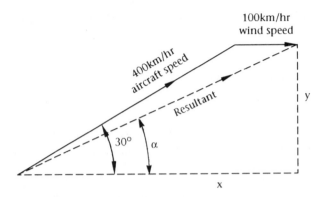

**Fig. 2.11**

easterly direction and parallel to the equator, is 100 km/h. Find the true velocity and direction of the aircraft (see Figure 2.11).

*Solution*

The vertical component $y$ and the horizontal component $x$ can be found using trigonometry:

$$y = 400 \sin 30° = 181.6 \, \text{km/h}$$
$$x = 100 + 400 \cos 30° = 456.4 \, \text{km/h}$$

Using Pythagoras' theorem to find the true velocity:

$$\text{true velocity} = \sqrt{(181.6^2 + 456.4^2)}$$
$$= 491.2 \, \text{km/h}$$

Using trigonometry to find angle $\alpha$:

$$\alpha = \sin^{-1}\left(\frac{181.6}{491.2}\right) = 24.1°$$

### 2.5.3 Resolution of velocity into two components

Just as two velocity vectors can be resolved into a single velocity vector, a single velocity vector can be split into two vectors as shown in Figure 2.12. Figure 2.12(a) shows a velocity vector AB. This can be resolved into two components AC and CB which lie at right angles to each other as shown in Figure 2.12(b). When splitting a vector into two it is usual to refer each new vector to the horizontal and vertical axes. These are designated as the $x$ and $y$ axes as shown in Figure 2.13.

Vectors which are resolved in such a way are called 'rectangular components' and are useful for solving a wide range of problems, particularly those involving trajectories.

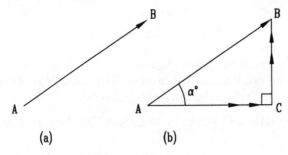

Fig. 2.12 Resolution of a single vector into two components.

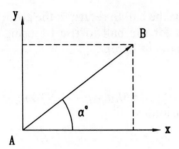

Fig. 2.13 Vector AB referred to right-angle axes.

## Example 2.10

If a ball is thrown at a velocity of 20 m/s at an angle of 60° to the horizontal (Figure 2.14), find:

(a) the horizontal and vertical components of the velocity;
(b) the total time for the ball to rise to the top of the trajectory and return to the ground;
(c) the horizontal distance travelled as the ball hits the ground.

Assume $g = 9.81 \, \text{m/s}^2$, the ground to be level and the ball to have negligible air resistance.

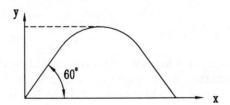

Fig. 2.14 Trajectory of ball.

*Solution*

For almost any trajectory problem most information is known about the vertical component. This must, therefore, be considered first. The link between the vertical and horizontal components is the total time for the ball to rise and fall back to the ground. The same time is taken for the movement of the ball in the horizontal direction.

(a) The horizontal and vertical components can be found using trigonometry:

$$v_{\text{VERT}} = 20 \sin 60° = 17.32 \, \text{m/s}$$
$$v_{\text{HORIZ}} = 20 \cos 60° = 10.0 \, \text{m/s}$$

(b) The total time for the ball to return to the ground can be found by first finding the time for the ball to rise by using equation (2.2) (ignore horizontal component here):

$$v_2 = v_1 + at$$

but $v_1 = 17.32 \, \text{m/s}$, $v_2 = 0$, $a = g = -9.81 \, \text{m/s}^2$ (note negative value denotes deceleration), so that

$$0 = 17.32 + (-9.81 \times t)$$

$$t = \frac{0 - 17.32}{-9.81} = 1.765 \, \text{s}$$

The time taken to descend is the same time as is taken to rise, i.e.

total time $t_{\text{TOT}} = 2 \times t = 2 \times 1.765 = 3.53 \, \text{s}$

(c) The total horizontal distance travelled can be found using equation (2.3)

$$s = \frac{(v_1 + v_2)t}{2}$$

Although the ball slows, stops and then accelerates back to earth in the vertical component, the horizontal component is independent and can, therefore, be treated as constant velocity. But $v_1 = v_2 = 10 \, \text{m/s}$, $t = 3.53 \, \text{s}$, so that

$$s = \frac{(10 + 10)3.53}{2} = 35.3 \, \text{m}$$

## 2.6 ANGULAR MOTION

Linear motion, described earlier, possesses qualities of displacement, velocity and acceleration. Rotating objects may also be described in these terms; however, special consideration needs to be given to the units involved. For instance, displacement becomes angular displacement, velocity becomes angular velocity and acceleration becomes angular acceleration. It is inappropriate to use metres as a unit of angular measure and SI does not allow the use of revolutions or degrees to measure angular motion. A unit designated the **radian** is used and allows uniformity of angular measurement no matter what the radius at which the action takes place. The concept of the radian is described below.

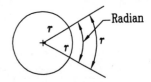

Fig. 2.15 The radian.

### 2.6.1 Angular displacement

A chalk mark drawn across the face of a bicycle tyre will move in a circular path when the wheel is rotated about its axis. The angle turned through by the chalk mark is the angular displacement, designated by the angle $\theta$ (theta).

The unit of angular displacement is the radian, which is a means of measuring an angle by taking account of the radius of the rotating object. If the radius of a circle were bent around the circumference, the angle subtended would be 1 radian, as shown in Figure 2.15.

The advantage of this is that no matter what the size of the circle, the angle subtended will always be the same. Consider a car wheel and a large tractor wheel. When the radius of each wheel is laid around its respective circumference, the angle thus marked out will be exactly the same. The concept of the radian as a measure of angular displacement can, therefore, be applied to any size of rotating component and still offer a uniform means of measuring the angular displacement.

### 2.6.2 Radians in a circle

The radius of a circle can be laid around the periphery of the circle as shown in Figure 2.16. It can be seen that the radius lies around the circumference six times with a small portion left over. This manœuvre can be done mathematically as follows:

$$\text{number of radians} = \frac{\text{circumference}}{\text{radius}} = \frac{2\pi r}{r} = 2\pi = 6.284$$

It should be noted that the radian is dimensionless.

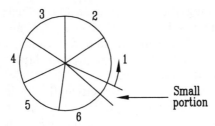

Fig. 2.16 Radians in a circle.

**Example 2.11**

If a point on a shaft is given an angular displacement of 1.2 radians, what is the angular displacement in degrees?

*Solution*

The angular displacement in degrees can be found by considering that 360 degrees is equivalent to $2\pi$ rad:

$$2\pi \, \text{rad} = 360°$$

$$1 \, \text{rad} = \frac{360°}{2\pi} = 57.3°$$

Therefore, angle turned through $= 1.2 \times 57.3° = 68.8°$.

**Example 2.12**

Calculate the linear displacement covered by a point on the tread of a car tyre of diameter 560 mm if it rotates through 1.5 radians.

*Solution*

The linear displacement can be found by considering that $2\pi$ rad represents a full circle:

$$1.5 \, \text{rad} = \frac{1.5}{2\pi} \, \text{parts of a circle}$$

total circumference $= 2\pi r$

$$= 2\pi \times \frac{560}{2} = 560\pi$$

Part of the circumference travelled

$$= \text{fraction of circumference} \times \text{total circumference}$$

$$= \frac{1.5 \times 560\pi}{2\pi} = 420 \, \text{mm}$$

**2.6.3 Angular velocity**

A car travelling along the road at a steady velocity possesses linear velocity measured in m/s. The wheels are also rotating at a steady velocity and their 'angular velocity' is measured in rad/s and is defined as 'the rate of change of angular displacement compared to elapsed time' and may be denoted by the symbol $\omega$ (omega).

Angular velocity parallels linear velocity in that it shares the same concepts of average velocity and constant velocity, but it must be noted that these now refer to rotational velocities and are defined in appropriate units.

If a chalk mark on a bicycle wheel takes a time $t$ to rotate through an angle $\theta$, then the average angular velocity is

$$\omega = \frac{\theta}{t} = \frac{\text{angular displacement}}{\text{time}} \tag{2.6}$$

Constant angular velocity means that equal, consecutive angular displacements (rad) are covered in equal intervals of time. The bicycle wheel would, therefore, rotate at a non-varying velocity.

Angular velocity is related to the 'frequency of rotation', or the number of revolutions made per second. Since there are $2\pi$ radians in one revolution, then $n$ revolutions per second will have an angular velocity of $2\pi n$ rad/s, as follows:

$$1 \, \text{rev/s} = 2\pi \ \text{rad/s}$$
$$2 \, \text{rev/s} = 2\pi \times 2 \ \text{rad/s}$$
$$3 \, \text{rev/s} = 2\pi \times 3 \ \text{rad/s}$$
$$n \, \text{rev/s} = 2\pi \times n \ \text{rad/s}$$

Hence

$$\text{angular velocity} = \omega = 2\pi n \, \text{rad/s} \tag{2.7}$$

In many cases angular velocity is quoted in the form of revolutions per minute, which requires a modification of equation (2.7), as follows:

$$\omega = \frac{2\pi n}{60} \, \text{rad/s} \tag{2.8}$$

where $n$ is expressed in rev/min.

### Example 2.13

A flywheel rotates at 3000 rev/min. Find its angular velocity in rad/s.

*Solution*

The angular velocity can be found by using equation (2.8):

$$\omega = \frac{2\pi n}{60} = \frac{2\pi \times 3000}{60} = 314 \, \text{rad/s}$$

### 2.6.4 Angular acceleration

When traffic lights turn green a car steadily builds up its velocity after being at rest. The car experiences linear acceleration. The wheels are also stationary when the lights change to green and they also steadily build up their velocity. Their angular velocity increases second by second and can be said to possess 'angular acceleration', denoted by $\alpha$ (alpha). The units of angular acceleration are radians/second/second (rad/s$^2$) and can be defined as 'the rate of change of angular velocity when compared to elapsed time'.

Angular acceleration may be considered in the same way as linear acceleration. For instance, if a rotating disc steadily changes its angular velocity from $\omega_1$ to $\omega_2$ in a time interval $t$, then the average angular acceleration is given by

$$\bar{\alpha} = \frac{(\omega_2 - \omega_1)}{t} \qquad (2.9)$$

where $\omega_1 =$ initial angular velocity (rad/s),
$\omega_2 =$ final angular velocity (rad/s),
$t =$ time interval (s).

### Example 2.14

The angular velocity of a grinding wheel changes from zero to 150 rad/s in 3 s. Find the average angular acceleration.

*Solution*

The angular acceleration can be found by using equation (2.9):

$$\bar{\alpha} = \frac{(\omega_2 - \omega_1)}{t} = \frac{(150 - 0)}{3} = 50 \,\text{rad/s}^2$$

### 2.6.5 Equations for angular motion

Angular motion is analogous to linear motion except that the variables $\theta, \omega$ and $\alpha$ are used to represent displacement, velocity and acceleration. Using these variables instead of $s$, $v$ and $a$ it is possible to modify the equations for linear motion so they are applicable to angular motion. Angular motion equivalent parameters need to be inserted as follows:

| Linear motion | | Angular motion | |
|---|---|---|---|
| $v = \dfrac{s}{t}$ | (2.1) | $\omega = \dfrac{\theta}{t}$ | (2.6) |
| $v_2 = v_1 + at$ | (2.2) | $\omega_2 = \omega_1 + \alpha t$ | (2.10) |
| $s = \dfrac{(v_1 + v_2)t}{2}$ | (2.3) | $\theta = \dfrac{(\omega_1 + \omega_2)t}{2}$ | (2.11) |
| $s = v_1 t + \frac{1}{2}at^2$ | (2.4) | $\theta = \omega_1 t + \frac{1}{2}\alpha t^2$ | (2.12) |
| $v_2^2 = v_1^2 + 2as$ | (2.5) | $\omega_2^2 = \omega_1^2 + 2\alpha\theta$ | (2.13) |

### Example 2.15

A wheel, initially at rest, is subjected to a constant angular acceleration of 2.5 rad/s$^2$ for 60 s. Find:

(a) the angular velocity attained;
(b) the number of revolutions made in that time.

*Solution*

(a) The angular velocity can be found using equation (2.10), but $\alpha = 2.5\,\text{rad/s}^2$, $t = 60\,\text{s}$, so

$$\omega_2 = \omega_1 + \alpha t$$
$$= 0 + (2.5 \times 60) = 150\,\text{rad/s}$$

(b) The angular displacement can be found using equation (2.12):

$$\theta = \omega_1 t + \tfrac{1}{2}\alpha t^2$$

so that

$$\theta = \frac{2.5 \times 60^2}{2} = 4500\,\text{rad}$$

To convert the angular displacement to revolutions it is recognized that one revolution represents $2\pi$ rad:

$$n = \frac{4500}{2\pi} = 716\,\text{rev}$$

## Example 2.16

A wheel initially has an angular velocity of 50 rad/s. When brakes are applied the wheel comes to rest in 25 s. Find the average retardation.

*Solution*

The angular retardation can be found using equation (2.10):

$$\omega_2 = \omega_1 + \alpha t$$

but $\omega_1 = 50\,\text{rad/s}$, $\omega_2 = 0$, $t = 25\,\text{s}$, so that

$$0 = 50 + (\alpha \times 25)$$

giving

$$\alpha = \frac{-50}{25} = -2\,\text{rad/s}^2$$

## Example 2.17

A drum starts from rest and attains a rotation of 210 rev/min in 6.2 s with uniform acceleration. A brake is then applied which brings the drum to rest in a further 5.5 s with uniform retardation. Find the total number of revolutions made by the drum.

*Solution*

The total number of revolutions can be found by using equation (2.11):

$$\theta = \frac{(\omega_1 + \omega_2)t}{2}$$

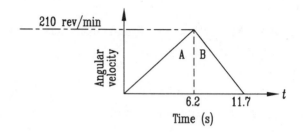

**Fig. 2.17**

Consider, initially, area A under the velocity–time graph (Figure 2.17). But $\omega_1 = 0, t = 6.2\,\text{s}$ and

$$\omega_2 = \frac{210 \times 2\pi}{60} = 22\,\text{rad/s}$$

so that

$$\theta = \frac{(0 + 22)6.2}{2} = 68.2\,\text{rad}$$

Consider area B under the velocity–time graph again using equation (2.11):

$$\theta = \frac{(\omega_1 + \omega_2)t}{2}$$

Now $\omega_1 = 22\,\text{rad/s}, \omega_2 = 0$ and $t = 5.5\,\text{s}$, so that

$$\theta = \frac{(22 + 0)5.5}{2} = 60.5\,\text{rad}$$

The total angular displacement $= 68.2 + 60.5 = 128.7\,\text{rad}$. Converting to revolutions

$$\theta_\text{rev} = \frac{128.7}{2\pi} = 20.5\,\text{rev}$$

### 2.6.6 Relationship between linear and angular motion

The common factor between linear and angular motion of, say, a wheel is the radius of the wheel. A larger radius will give a larger circumference and so the wheel will have to roll further along the ground before an angular displacement of 1 rad has been covered. This is demonstrated in Figure 2.18, which shows different sized wheels moving through the same angular displacement.

Though a larger distance is covered by a larger wheel, the radius is a unifying factor allowing the angle turned through to be related to the linear displacement.

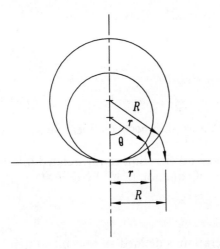

**Fig. 2.18** The relationship between linear and angular displacement.

The arc length $r$ (or $R$ for the larger circle) in Figure 2.18 is the radius of the circle which has been laid around the periphery. As the circle rolls along the flat surface the length of arc can be measured on the flat surface. The arc length represents the angular displacement of the circle and if the length of the arc can be found, the linear displacement will also be found. It is important to note that even though the radii of the circles are different and the linear displacement of the larger circle is greater than that of the smaller circle, the angle turned through is exactly the same for both circles and is one radian.

It follows that if the circle is rolled along the surface through an angle of one radian, a linear displacement equivalent to one radius will be covered.

The linear displacement $s$ can be found as follows:

$$s = \theta r \qquad (2.14)$$

where $\theta$ = angular displacement (rad),
$r$ = radius of wheel (m).

The link between linear velocity and angular velocity may be made by using equation (2.1):

$$v = \frac{s}{t}$$

But from equation (2.14)

$$s = \theta r$$

and

$$v = \frac{\theta r}{t}$$

But angular velocity is defined as

$$\omega = \frac{\theta}{t}$$

so that

$$v = \omega r \tag{2.15}$$

and presents the relationship between linear velocity and angular velocity.

Angular and linear acceleration can be related in a similar manner by dividing both sides of equation (2.15) by time $t$:

$$\frac{v}{t} = \frac{\omega r}{t}$$

but angular acceleration is defined as $\alpha = \omega/t$, so that

$$a = \alpha r \tag{2.16}$$

and represents the relationship between angular and linear acceleration.

### Example 2.18

A motor car with wheels of 0.6 m diameter accelerates uniformly from 3 to 18 m/s in 10 s. Find:

(a) the angular acceleration of the wheels;
(b) the number of revolutions made by each wheel during the speed change.

*Solution*

(a) The initial angular velocity and the final angular velocity can be found using equation (2.15):

$$v = \omega r$$

or

$$\omega = \frac{v}{r}$$

so that

$$\omega_1 = \frac{3}{0.3} = 10 \, \text{rad/s}$$

and

$$\omega_2 = \frac{18}{0.3} = 60 \, \text{rad/s}$$

The angular acceleration can be found using equation (2.10):

$$\omega_2 = \omega_1 + \alpha t$$

but $\omega_2 = 60\,\text{rad/s}$, $\omega_1 = 10\,\text{rad/s}$ and $t = 10\,\text{s}$, so that

$$60 = 10 + (\alpha \times 10)$$

and

$$\alpha = 5\,\text{rad/s}^2$$

(b) The number of revolutions turned through during the acceleration can be found using equation (2.11):

$$\theta = \frac{(\omega_1 + \omega_2)t}{2}$$

but $\omega_2 = 60\,\text{rad/s}$, $\omega_1 = 10\,\text{rad/s}$ and $t = 10\,\text{s}$, so that

$$\theta = \frac{(10 + 60)10}{2} = 350\,\text{rad}$$

Hence the number of revolutions turned through is

$$n = \frac{350}{2\pi} = 55.7\,\text{rev}$$

## 2.7 SUMMARY

Key equations which have been introduced in this chapter are as follows.

### 2.7.1 Equations for linear and angular motion

| Linear motion | | Angular motion | |
|---|---|---|---|
| $v = \dfrac{s}{t}$ | (2.1) | $\omega = \dfrac{\theta}{t}$ | (2.6) |
| $v_2 = v_1 + at$ | (2.2) | $\omega_2 = \omega_1 + \alpha t$ | (2.10) |
| $s = \dfrac{(v_1 + v_2)t}{2}$ | (2.3) | $\theta = \dfrac{(\omega_1 + \omega_2)t}{2}$ | (2.11) |
| $s = v_1 t + \frac{1}{2}at^2$ | (2.4) | $\theta = \omega_1 t + \frac{1}{2}\alpha t^2$ | (2.12) |
| $v_2^2 = v_1^2 + 2as$ | (2.5) | $\omega_2^2 = \omega_1^2 + 2\alpha\theta$ | (2.13) |

### 2.7.2 Relationships between linear and angular motion

| | |
|---|---|
| $s = \theta r$ | (2.14) |
| $v = \omega r$ | (2.15) |
| $a = \alpha r$ | (2.16) |
| $\omega = 2\pi n$ where $n = \text{rev/s}$ | (2.7) |
| $\omega = \dfrac{2\pi n}{60}$ where $n = \text{rev/min}$ | (2.8) |

## 2.8 PROBLEMS

### 2.8.1 Linear motion

1. A person walks a distance of 8 km in a direction 20° east of north and then walks another 6 km in a direction 50° south of east. Using vector diagrams find the magnitude and direction of the final displacement relative to the starting point.
2. A train accelerates uniformly from rest to reach 54 km/h in 200 s, after which the speed remains constant for 300 s. At the end of this time the train decelerates to rest in 150 s. Find the total distance travelled.
3. The cutting stroke of a planing machine is 600 mm and it is completed in 1.2 s. For the first and last quarters of the stroke the table is uniformly accelerated and retarded, the speed remaining constant during the remainder of the stroke. Determine the maximum cutting speed.
4. The following table shows how the velocity of a car varied with time:

| Velocity (km/h) | 0 | 19 | 35 | 48 | 48 | 27 | 8 |
|---|---|---|---|---|---|---|---|
| Time (s) | 0 | 3 | 6 | 9 | 12 | 15 | 18 |

Assuming these speed values to be joined by straight lines, plot a graph of speed against time and determine:
   (a) the distance travelled;
   (b) the acceleration during the first 3 s;
   (c) the retardation during the last 3 s;
   (d) the average speed in km/h.
5. In the final stages of a moon landing a lunar module descends under the thrust of its descent engine to within 5 m of the surface, where it has a downward velocity of 4 m/s (Figure 2.19). If the descent engine is abruptly cut off at this point, find the impact velocity on the surface. Take the lunar gravitational constant to be 1.64 m/s².
6. A cannon-ball is fired from the top of a cliff with a horizontal muzzle velocity of 120 m/s (Figure 2.20). If the cannon is 60 m above the ground, find (a) the time for the cannon-ball to hit the ground and (b) the range.

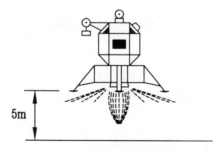

5m

**Fig. 2.19**

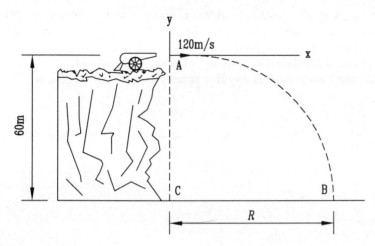

**Fig. 2.20**

## 2.8.2   Angular motion

7. The wheels of a car increase their rate of rotation from 1.0 to 8.0 rev/s in 20 s. Find:
   (a) the angular acceleration of the wheels;
   (b) the linear acceleration of a point on the rim of a wheel if the radius is 350 mm.
8. The speed of a shaft increases from 300 to 360 rev/min while turning through 18 complete revolutions. Find:
   (a) the angular acceleration;
   (b) the time taken for this change.
9. After the power to a drive shaft is cut off, it describes 120 revolutions in the first 30 s and finally comes to rest during a further 30 s. If the retardation is uniform, find:
   (a) the initial angular velocity in rev/min;
   (b) the retardation in rad/s.
10. A swing bridge has to be turned through 90° in 140 s. The first 60 s is a period of uniform acceleration. The subsequent 40 s is a period of uniform velocity and the final period of 40 s is of uniform angular retardation. Find:
    (a) the maximum angular velocity;
    (b) the angular acceleration;
    (c) the angular retardation.
11. The armature of an electric motor rotating at 1500 rev/min has a brake applied to it which retards the motion uniformly at 1.5 rad/s until the speed is reduced to 750 rev/min. The speed is then maintained for 30 s, after which the armature is uniformly accelerated for 20 s until the speed again reaches 1500 rev/min. Find:
    (a) the total time taken;
    (b) the total number of revolutions made by the armature.

12. At a particular instant a trolley is moving with a velocity of 5.0 m/s and a linear acceleration of 0.75 m/s$^2$. The trolley is carried on wheels of diameter 50 mm. Find:

(a) the instantaneous angular velocity of the wheels;

(b) the instantaneous angular acceleration of the wheels.

# Newton's laws, impulse and momentum

**3**

## 3.1 AIMS

- To explain the interrelationship between motion, mass, inertia and forces according to the rules set out in Newton's laws.
- To define the equations which can be used to apply Newton's laws to both linear and angular motion.
- To introduce and define impulse and momentum, in both linear and angular terms.
- To introduce and explain the principle of conservation of momentum.
- To introduce a method of analysis by which problems relating to Newton's Laws may be solved.

## 3.2 NEWTON'S LAWS

Sir Isaac Newton first laid the foundations for what is known as Newtonian mechanics, by observing how masses can be manipulated by applied forces. The rules he developed are known as Newton's laws and define the interactions between masses, forces, velocity and acceleration and have already been briefly mentioned in Chapter 1.

Newton also realized that all masses possess the 'quality of inertia', which requires forces to be used to move a mass from rest, or stop it, or reduce its velocity. Inertia cannot be felt, as can mass; however, it always accompanies mass and its effect is to oppose any force which attempts to change the state of mass.

The effects of inertia are fundamental to Newton's laws, both in linear and angular forms. However, it is in consideration of the angular forms of Newton's laws that inertia is used as a calculated, finite quantity to describe the distribution of mass about an axis. For example, a bicycle wheel will possess a larger value of inertia than the same mass concentrated in a small diameter rotor. This is because most of the mass of the bicycle wheel is placed at the outer rim. It is true to say that the larger the radius at which the mass is 'placed', the greater the value of inertia.

### 3.2.1 Newton's first law

A body will remain at rest or continue with a uniform velocity in a straight line, unless acted upon by an external force.

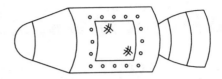

**Fig. 3.1** Interstellar spacecraft.

Perhaps this law should be called 'The law of inertia' since it is the mass, or the inertia which is possessed by the mass, which compels it to stay at rest or continue to move unless influenced by some external force. Consider the intersteller spacecraft (ISC) shown in Figure 3.1. It is stationary in space after transferring cargo from a shuttle craft. The ISC will remain stationary until it is acted upon by an external force in the form of thrust from the rocket motors. The rocket motors will apply a force, accelerating the ISC until it is travelling at an appropriate interstellar velocity, at which point they will be switched off. The inertia possessed by the ISC will ensure it continues at a constant velocity and since there is no air resistance in space, it will continue at that velocity.

In due course the spacecraft approaches its destination planet and needs to reduce velocity to take up orbit. The forward thrusters apply an external force against the motion, thus decelerating the spacecraft.

The example of the spacecraft illustrates that the application of an external force will increase the velocity of the mass, imparting momentum by virtue of the inertia possessed by the mass. The momentum ensures that the spacecraft continues at a constant velocity. The application of an external force in the opposite direction reduces its forward speed and thus destroys its forward momentum.

Since the ISC is travelling at a constant velocity during its journey it can be said to possess 'momentum' which can be described as:

$$\text{momentum} = \text{mass} \times \text{velocity}$$

or

$$\text{momentum} = m \times v \tag{3.1}$$

and has units of kg m/s.

In consideration of equation (3.1), the mass of the spacecraft will not change. As velocity increases at the start of the journey, so does momentum. During the journey velocity and, hence, momentum remain constant. This does not change until the spacecraft slows. In reducing its velocity, momentum is also reduced.

### 3.2.2  Newton's second law

When a body is acted upon by an external force the rate of change of momentum is directly proportional to the applied force and in the same direction as that force.

At the start of the journey, the interstellar spacecraft, in Figure 3.1, is stationary and is acted upon by an external force in the form of the rocket motors which accelerate the craft. After the rocket motors are cut off, the ISC will continue at a constant velocity until the thrusters are employed to decelerate the craft in preparation for orbit around the destination planet. Deceleration may be considered as negative acceleration.

Whenever an external force is applied the state of momentum is changed. At the start of the journey momentum is increased from zero due to the acceleration, while at the end of the journey the ISC is slowed by the thrusters, thus reducing momentum. These changes in momentum are directly proportional to the forces applied.

Since the applied force causes the acceleration or deceleration of the mass, the following equation may be applied:

$$F = ma \tag{3.2}$$

where  $F = $ force (N),
$\quad m = $ mass (kg),
$\quad a = $ acceleration (m/s$^2$).
This can be alternatively considered as follows:

$$F = m \times \frac{\Delta v}{\Delta t}$$

where $\Delta v / \Delta t$ is the rate of change of velocity with respect to time. If the mass of the body is constant, this can be re-expressed as

$$F = \frac{\Delta(mv)}{\Delta t}$$

which is the form given in equation (1.2).

### 3.2.3  Newton's third law

To every force there is an equal and opposite reaction and this reaction is always colinear.

When the rocket motors of the spacecraft provide thrust, the direction of the motion is in the same direction as the thrust. In order to increase the velocity of the spacecraft, the motors must provide a force equal and opposite to the resistance to motion offered by the inertia possessed by the mass. The action and reaction occur whether the force is applied while the spacecraft is stationary or whether it is moving.

The same principle can be applied to a beam applying a downward force on a support. The support must apply an exactly equal and opposite reactive force to that force applied by the beam. The system can be said to be in equilibrium and is stationary since the downward forces exactly equal the upward forces.

Consider, now, a gun firing a shell. The force which creates the recoil of the gun is exactly opposite and equal to the force which acts on the shell.

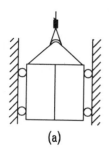

(a)

**Fig. 3.2(a)**

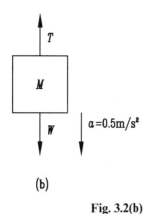

(b)

**Fig. 3.2(b)**

### Example 3.1

An elevator and passengers have a mass of 1750 kg (Figure 3.2(a)). Calculate the tension in the suspending cable when the elevator is:

(a) stationary
(b) descending with an acceleration of 0.5 m/s².

*Solution*

(a) The force in the cable when the elevator is stationary can be found using equation (3.2):

$$F = ma$$

Now $m = 1750$ kg, $a = g = 9.81$ m/s², so that

$$F = 1750 \times 9.81 = 17\,168\,\text{N} = 17.2\,\text{kN}$$

(b) Consider the free-body diagram (Figure 3.2(b)). The tension in the cable acts upwards and the weight of the elevator and passengers acts downwards. The forces may be summed and applied to equation (3.2) as follows.

Now $F = 17\,168$N, $m = 1750$ kg and $a = 0.5$ m/s².

$$F = ma$$
$$W - T = ma$$
$$g \times 1750 - T = 1750 \times 0.5$$
$$17\,168 - T = 875$$
$$T = 17\,168 - 875 = 16\,293\,\text{N} \quad \text{or} \quad 16.3\,\text{kN}$$

## 3.3 NEWTON'S LAWS APPLIED TO ROTATING BODIES

Newton's laws can be applied to rotating bodies in a similar manner to bodies moving with linear motion. In order to accommodate the requirements for rotary motion some basic angular concepts first need to be considered.

### 3.3.1 Moment

If a force is applied at a radius from an axis of rotation, as shown in Figure 3.3, the force will follow the periphery of the circle fixed by the radius.

The leverage which the force applies is dependent upon the radius at which the force acts. This leverage is called the moment and can be calculated as

$$\text{moment} = \text{force} \times \text{radius}$$

$$M = F \times r \tag{3.3}$$

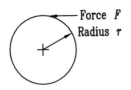

**Fig. 3.3** Force applied at a radius.

The units of moment are N m.

When the force is applied and the body remains static, as in the case of a beam, the action is called a moment. If the force causes rotation, as in the turning of a shaft or wheel, then the action is called a torque and can be calculated as

torque = force × radius

$$T = F \times r \qquad\qquad (3.4)$$

The units of torque are N m.

It should be noted that in both cases the force is applied at 90° to the line of the radius. Put another way, the force is always applied at a tangent to the periphery of the circle. Furthermore, the greater the radius, the greater the moment or torque.

**Example 3.2**

A force of 10 N is applied at a tangent to the periphery of a wheel whose radius is 0.05 m (Figure 3.4). The same force is then applied in the same manner to a wheel of 0.5 m radius. Find:

(a) the torque applied to the smaller wheel;
(b) the torque applied to the larger wheel.

*Solution*

(a) The torque applied to the smaller wheel can be found using equation (3.4):

$T = Fr$

Now $F = 10$ N and $r = 0.05$ m, so that

$T = 10 \times 0.05 = 0.5$ N m

(b) The torque applied to the large wheel can be found using equation (3.4):

$T = Fr$

Now $F = 10$ N and $r = 0.5$ m, so that

$T = 10 \times 0.5 = 5$ N m

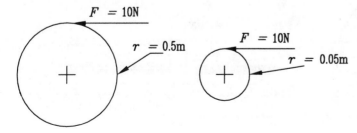

Fig. 3.4

By comparing the above answers it is clear that the torque increase is solely due to the increase in the radius.

Torque, then, can be considered to be the angular equivalent of force since it is the product of force and the radius at which it acts.

### 3.3.2 Newton's second law applied to an angular motion

It has previously been shown that Newton's second law utilizes equation (3.2) when applied to linear motion:

$$F = ma$$

In order to use Newton's second law for angular motion, angular equivalents for force, mass and acceleration may be used:

$$T = I\alpha \tag{3.5}$$

where $I$ = moment of inertia discussed in section 3.3.3.

A distinction should be drawn between equation (3.4)

$$T = Fr$$

and equation (3.5)

$$T = I\alpha$$

Equation (3.4) describes the torque required to turn, say, a handle against an opposing torque, while equation (3.5) describes the torque required to overcome the inertia of a flywheel to increase the speed of rotation.

### 3.3.3 Determination of the moment of inertia

Rotors and wheels with large moments of inertia have, for many years, been used by engineers as a mechanical means of storing energy. Perhaps the most distinctive of these are the flywheels used with steam engines, one of which is illustrated in Figure 3.5.

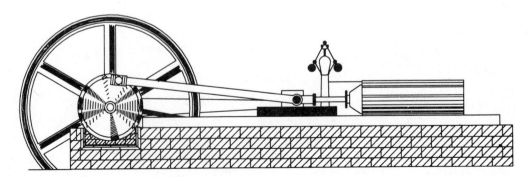

**Fig. 3.5** Line diagram of a steam engine.

Figure 3.5 shows a flywheel that has most of its mass placed at the greatest radius and so it possesses the largest value of the moment of inertia possible with the mass available. The function of the flywheel is twofold. For a single acting steam engine, upon start-up, the piston would power the mechanism half-way round the cycle and since there is no return power stroke the mechanism would stall. A flywheel connected to the crankshaft would absorb some of the energy from the power stroke of the piston, then use it to return the piston to the start of the cycle.

At speed, the flywheel absorbs and releases energy at a much faster rate and prevents the mechanism from 'pulsing' in time with the steam injected at the cylinder. The presence of the flywheel smooths out the pulses to give a more uniform speed. This is also true of a car engine having several cylinders.

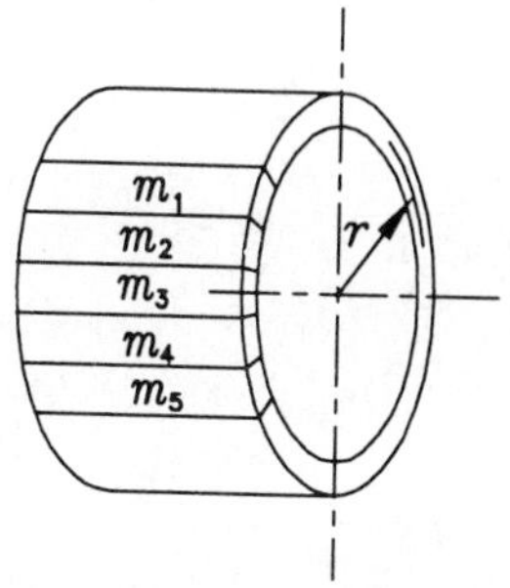

**Fig. 3.6** A rim-type flywheel split into segments.

The analysis of a rim-type flywheel shows how the moment of inertia may be calculated.

Figure 3.6 shows the rim of a flywheel split into equal segments. The moment of inertia $I$ can be calculated according to

$$I = \sum mr^2$$

or

$$I = m_1 r_1^2 + m_2 r_2^2 + m_3 r_3^2 + m_4 r_4^2 \dots \text{ etc.} \tag{3.6}$$

Since $r_1$, $r_2$, $r_3$ and $r_4$ are the same value the equation can be written

$$I = (m_1 + m_2 + m_3 + m_4)r^2$$

and gives

$$I = (\text{the total mass of the ring})r^2$$

or

$$I = mr^2 \tag{3.7}$$

This assumes that the total mass of the ring acts at the radius $r$.

## Example 3.3

Figure 3.7 shows a steel flywheel with the mass concentrated at the rim. Determine:

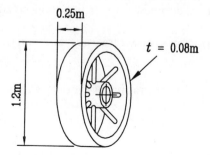

**Fig. 3.7**

(a) the mass of the flywheel assuming that the hub and spokes have no significant contribution and that the density of steel is $7800 \, \text{kg/m}^3$;

(b) the moment of inertia.

*Solution*

(a) The mass can be found as follows:

$$\text{volume} = 2\pi r \times \text{width} \times \text{thickness}$$

and

$$\text{mass} = 2\pi r \times \text{width} \times \text{thickness} \times \text{density}$$

Now width $= 0.25 \, \text{m}$, thickness $= 0.08 \, \text{m}$, density $= 7800 \, \text{kg/m}^3$ and mean radius is

$$r = \frac{1.2 - 0.08}{2} = 0.56$$

so that

$$m = 2\pi \times 0.56 \times 0.25 \times 0.08 \times 7800$$
$$= 549 \, \text{kg}$$

(b) The moment of inertia can be found using equation (3.7):

$$I = mr^2 = 549 \times 0.56^2 = 172 \, \text{kg} \, \text{m}^2$$

### 3.3.4 Radius of gyration

The examples covered so far have assumed that the mass can be taken to act at a specific radius and that the mass is concentrated at that radius. Many engineering components possess mass which is distributed between the centre and the outer radius. Such a component is the disc rotor, whose mass is distributed between the centre of the disc and its periphery. Clearly the concept of mass acting at a discrete radius needs to be modified.

Instead of the radius $r$, an average radius is used, called the 'radius of gyration' and designated $k$. This is the special radius which can be considered to be the radius at which a concentrated mass would have to be situated so that the moment of inertia would be equal to that of the whole body.

The radius of gyration $k$ now replaces $r$ and equation (3.7) becomes

$$I = mk^2 \tag{3.8}$$

Some basic examples of the radius of gyration can be seen in Figure 3.8.

### Example 3.4

An angular acceleration of $10 \, \text{rad/s}^2$ is to be imparted to a solid disc with a peripheral radius of $0.3 \, \text{m}$ and mass of $30 \, \text{kg}$ (Figure 3.9). Find:

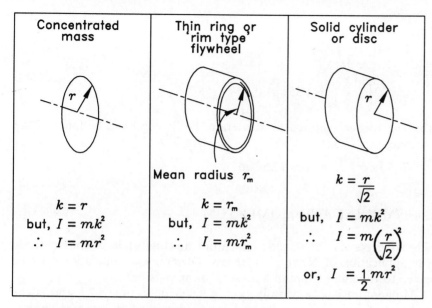

Fig. 3.8 Examples of the radius of gyration.

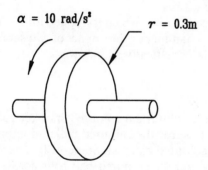

Fig. 3.9

(a) the moment of inertia of the flywheel;
(b) the torque required to accelerate the flywheel.

*Solution*

(a) The moment of inertia of the flywheel may be found using equation (3.8):

$$I = mk^2$$

Inserting

$$k = \frac{r}{\sqrt{2}}$$

gives

$$I = m\left(\frac{r}{\sqrt{2}}\right)^2$$

$$= 30\left(\frac{0.3}{\sqrt{2}}\right)^2 = 1.35 \, \text{kg m}^2$$

(b) The torque required to accelerate the flywheel can be found using equation (3.5):

$$T = I\alpha = 1.35 \times 10 = 13.5 \, \text{Nm}$$

## 3.4  IMPULSE AND MOMENTUM

The concepts of impulse and momentum are closely related and are really the application of Newton's first law. Generally consideration may be given to the momentum of a mass or an impulsive force.

Momentum has previously been defined in section 3.2.1 and is the product of mass and velocity and can be described by equation (3.1):

$$\text{momentum} = mv$$

where the units are kg m/s.

An impulsive force is a force which is applied for a short period of time, such as a blow with a hammer or the impact of two snooker balls. Impulse can be described mathematically as follows:

$$\text{impulse} = F \times t \tag{3.9}$$

where the units are N s.

Consider a white ball hitting a coloured snooker ball. In certain shots the white ball rolls towards the coloured ball and upon impact passes all of its energy to the coloured ball instantaneously. The white ball becomes instantly stationary and the coloured ball rolls across the table.

Initially, the white ball possesses momentum since it has mass and velocity. At the instant of impact the momentum is imparted to the coloured ball by the impulsive nature of the impact and this can be described in terms of $Ft$.

It follows that, since all the momentum has been transferred from one ball to the other by the impact,

$$\text{impulse of the force} = \text{momentum}$$

$$Ft = mv \tag{3.10}$$

If the coloured ball is already moving when the impact occurs, the momentum which it already possessed will be increased. Indeed the white ball may not transfer all its momentum and will continue rolling at a slower pace. The loss of momentum of the white ball is equal to the gain in momentum of the coloured ball and is transferred due to the mechanism of the impact. Since no mass is lost or gained by the momentum transfer, the change in momentum must be directly proportional to the change in velocity.

Equation (3.10) can be modified to account for the changes in momentum experienced by a mass, such as the white ball or the coloured ball, no matter what the state of its motion:

impulse = change in momentum

$$Ft = mv_2 - mv_1 \tag{3.11}$$

### 3.4.1 Conservation of momentum

Consider the interstellar spacecraft mentioned earlier. Its momentum carries it through space at a constant velocity and unless acted upon by an external force, such as rocket motors, it will continue at that velocity. If the rocket motors were used to slow the spacecraft, then some or all of the momentum would be destroyed by the external force.

From this it can be deduced that 'momentum can only be destroyed by a force or created by the action of a force'. If no external force is exerted on the body, then the momentum remains constant.

Another way of describing the conservation of momentum is: 'the total momentum of a system before impact is equal to the total momentum after impact'. The impulsive force at impact destroys all or some of the momentum of one body and imparts it to the second body. The total value of the momentum always remains constant. It is merely shared between the masses involved. A white ball and a coloured snooker ball, both travelling across a table, possess individual values of momentum. After their impact the individual velocities will have changed, imparting different values of momentum to each ball; however, the total momentum will have remained the same in any direction.

It can be said that the total momentum before impact is equal to that after impact, so that

total momentum before impact = total momentum after impact

$$m_1v_1 + m_2v_2 = m_1v_3 + m_2v_4 \tag{3.12}$$

### Example 3.5

A railway wagon, of mass 40 t travels along a track at 20 km/h, and collides with another wagon, of mass 15 t, travelling in the opposite direction, at 25 km/h (Figure 3.10). After impact the first travels in the original direction at a velocity of 5 km/h. Find the velocity of the second wagon after the impact.

*Solution*

The velocity of the second wagon can be found using equation (3.12):

$$m_1v_i + m_2v_2 = m_1v_3 + m_2v_4$$

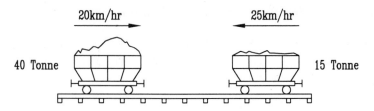

**Fig. 3.10**

Now $m_1 = 40\,\text{t}$, $v_1 = 20\,\text{km/h}$, $m_2 = 15\,\text{t}$, $v_2 = 25\,\text{km/h}$ and $v_3 = 5\,\text{km/h}$, so that

$$(40 \times 20) + (15 \times -25) = (40 \times 5) + (15 \times v_4)$$

$$v_4 = \frac{800 - 375 - 200}{15} = 15\,\text{km/h}$$

*Note:* while velocities are generally quoted in m/s, km/h or any other consistent units may be used.

### Example 3.6

A shell is fired from a gun at 500 m/s at an angle of 30° to the horizontal (Figure 3.11). If the mass of the shell is 10 kg and the velocity of recoil of the gun is to be limited to 1.5 m/s, find the required minimum mass of the gun.

*Solution*

The mass of the gun may be found by using equation (3.1):

$$\text{momentum} = mv$$

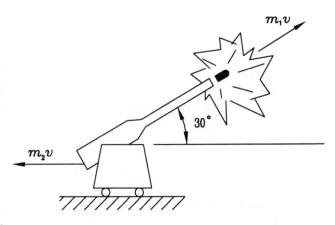

**Fig. 3.11**

and since the momentum of the shell must equal the momentum of the gun and carriage, this can be modified as follows:

gun momentum = shell momentum parallel to track

$$m_G v_G = (m_s v_s) \cos 30°$$

so that

$$m_G \times 1.5 = (10 \times 500) \cos 30°$$

$$m_G = \frac{4330}{1.5} = 2887 \, kg$$

### 3.4.2  Angular momentum

Momentum has previously been discussed in terms of linear motion. Momentum concerning rotating masses can be quantified, however, and is an important part of engineering analysis. The concepts of impulse and momentum which apply to linear motion also apply to angular motion and it is merely a case of inserting angular equivalents into the equations already developed, as follows:

Linear momentum

$$\text{momentum} = mv \tag{3.1}$$

$$\text{impulse} = Ft \tag{3.9}$$

$$Ft = mv \tag{3.10}$$

$$Ft = mv_2 - mv_1 \tag{3.11}$$

$$m_1 v_1 + m_2 v_2 = m_1 v_3 + m_2 v_4 \tag{3.12}$$

Moment of momentum

$$\text{angular momentum} = I\omega \tag{3.13}$$

$$\text{angular impulse} = Tt \tag{3.14}$$

$$Tt = I\omega \tag{3.15}$$

$$Tt = I\omega_2 - I\omega_1 \tag{3.16}$$

$$I_1 \omega_1 + I_2 \omega_2 = I_1 \omega_3 + I_2 \omega_4 \tag{3.17}$$

The moment of inertia can be found in the way described in section 3.3.4 for solid rotors. However, for multiple discrete masses fixed to the same shaft, a summation method must be used as follows. Since the angular momentum = $I\omega$ and $I = mr^2$ so that angular momentum = $mr^2 \omega$, for multiple discrete masses

$$\text{angular momentum} = (m_1 r_1^2 + m_2 r_2^2 + m_3 r_3^2 + \ldots)\omega$$

where $\omega$ is the common angular velocity, giving

$$\text{angular momentum} = \sum (mr^2)\omega \quad \text{or} \quad = \sum I\omega$$

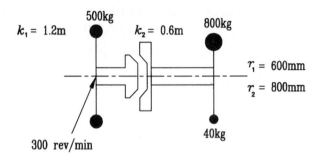

**Fig. 3.12**

**Example 3.7**

A flywheel and shaft of 500 kg total mass and 1.2 m radius of gyration are rotating in fixed bearings at 300 rev/min (Figure 3.12). By means of a clutch, another similar flywheel system is suddenly connected to the first shaft. The second shaft is made up of two discrete masses of 800 and 40 kg. The radii at which these two masses can be taken to act are 600 and 800 mm respectively. Find:

(a) the moment of inertia for each shaft/flywheel system;
(b) the new common speed of rotation if the second shaft/flywheel system is initially at rest.

*Solution*

(a) The moment of inertia for the first shaft/flywheel system can be found using equation (3.8):

$$I = mk^2 = 500 \times 1.2^2 = 720\,\text{kg m}^2$$

The moment of inertia for the second shaft/flywheel system can be found using equation (3.6):

$$I_2 = m_1 r_1^2 + m_2 r_2^2$$
$$= (800 \times 0.6^2) + (40 \times 0.8^2) = 313.6\,\text{kg m}^2$$

(b) The common speed of the two connected shaft/flywheel systems can be found using equation (3.17):

$$I\omega_1 + I_2\omega_2 = I_1\omega_3 + I_2\omega_4$$

where

$$\omega_1 = \frac{300 \times 2\pi}{60} = 31.4\,\text{rad/s}$$

and $\omega_3 = \omega_4$, so that

$$I_1\omega_1 + I_2\omega_2 = (I_1 + I_2)\omega_3$$
$$(720 \times 31.4) + (313.6 \times 0) = (720 + 313.6)\omega_3$$
$$22608 + 0 = 1033.6\omega_3$$
$$\omega_3 = 21.9\,\text{rad/s}$$

or

$$\frac{21.9 \times 60}{2\pi} = 208.9\,\text{rev/min}$$

## 3.5   SUMMARY

The key equations introduced in this chapter are:

Linear applications

| | |
|---|---|
| momentum $= mv$ | (3.1) |
| $F = ma$ | (3.2) |
| impulse $= Ft$ | (3.9) |
| $Ft = mv$ | (3.10) |
| $Ft = mv_2 - mv_1$ | (3.11) |
| $m_1v_1 + m_2v_2 = m_1v_3 + m_2v_4$ | (3.12) |

Angular applications

| | |
|---|---|
| angular momentum $= I\omega$ | (3.13) |
| $T = I\alpha$ | (3.5) |
| angular impulse $= Tt$ | (3.14) |
| $Tt = I\omega$ | (3.15) |
| $Tt = I\omega_2 - I\omega_1$ | (3.16) |
| $I_1\omega_1 + I_2\omega_2 = I_1\omega_3 + I_2\omega_4$ | (3.17) |

## 3.6   PROBLEMS

1. A 3000 kg lorry is travelling at 40 km/h on a level road. If it is brought to rest with uniform retardation in 15 s, find:
   (a) the retarding force
   (b) the distance travelled.
2. Find the force exerted on the floor of a lift by a person having a mass of 90 kg, when the lift is:
   (a) ascending with an acceleration of 2.0 m/s²;
   (b) descending with an acceleration of 2.5 m/s².

3. A railway wagon having a mass of 22 000 kg, is travelling at 30 km/h along a straight, level track. The brakes are applied, providing a uniform retardation of 2.5 m/s². Find:
   (a) the distance travelled by the wagon before coming to rest;
   (b) the retarding force.

4. A flywheel has a radius of gyration $k$ of 0.5 m and a mass of 400 kg. Find:
   (a) the moment of inertia;
   (b) the constant torque needed to bring such a wheel to rest in 40 s, if it were originally rotating at 6.0 rev/s.

5. A flywheel has a mass of 70 kg and a radius of gyration of 400 mm. Find the torque needed to increase the speed from 6 to 8 rev/s in 40 s.

6. A pulley whose mass is 50 kg and whose radius of gyration is 0.75 m has the following masses attached to it: 5.0 kg at 60 mm radius, 7 kg at 600 mm radius, 900 kg at 300 mm radius and 12 kg at 750 mm radius. Find:
   (a) the total radius of gyration of the complete system;
   (b) the torque required to accelerate the system from 0 to 150 rev/s in 90 s.

7. A railway truck of 9 t travelling due east with a velocity of 12 km/h collides with a second railway truck of mass 14 t travelling due west at a velocity of 4 km/h. Find the magnitude and direction of their common velocity if the two trucks are locked together after impact.

8. A rocket travelling in a horizontal straight line at 12 000 m/s explodes and breaks up into two parts of mass 100 and 300 kg. Both parts move in the same direction as before but the lighter section moves at a velocity 1000 m/s faster than the heavier part. Find the final velocity of each section.

9. An engine is suddenly coupled to a rotating drum by a friction clutch. The moment of inertia of the engine is equivalent to a mass of 25 kg acting with a radius of gyration of 393 mm and rotates at 200 rev/min before engagement. The drum has a mass of 60 kg and a radius of gyration of 406 mm and is initially at rest. Find the speed of the engine and drum immediately after connection.

10. A flywheel and shaft have a total mass of 516 kg and radius of gyration of 1.07 m and rotate at 150 rev/min. A clutch suddenly couples the flywheel and shaft to a machine whose rotating parts have a mass of 1670 kg and a radius of gyration of 0.623 m. If, after coupling, the speed of rotation is 60 rev/min, in the same direction as that of the flywheel initially, find the original speed of the rotating parts of the machine.

# Statics 4

## 4.1 AIMS

- To introduce forces and their application.
- To describe the types of force which may be encountered.
- To explain how forces and force systems may be analysed using vectors.
- To introduce the concept of equilibrium.
- To show how the concept of equilibrium can be used to analyse force systems containing two or more forces.
- To explain how forces acting on bodies or systems create moments and couples.

## 4.2 STATICS

Statics is the analysis of stationary objects under the influence of forces of various types. From Newton's first law of motion a body can remain stationary only if there is no force acting on the body. Alternatively, if there are several forces which balance each other, the body is in equilibrium and remains stationary. It is necessary to define the types of forces which may be encountered and examine the ways in which they may be analysed.

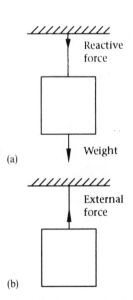

(a)

(b)

**Fig. 4.1** Suspended block showing (a) external forces, (b) reactive forces.

### 4.2.1 Applied forces

Applied forces are applied from an external source, such as when a block is suspended by a rope. The upward force in the rope is the applied force on the block, as shown in Figure 4.1(a). If the applied force due to the rope did not exist, the block would simply fall under the action of gravity.

### 4.2.2 Reactive forces

Newton's third law says that for every applied force there is a reactive force present. In Figure 4.1(b) the reactive force is the force induced in the rope by the weight of the block. Applied forces and reactive forces always act in pairs. One cannot exist without the other. They always act along the same line and in opposite directions.

### 4.2.3 Internal forces

Internal forces are the equal and opposite reactions to the externally applied loads. In Figure 4.2(a) the external forces are those which are applied by the weight of the trap-door to the cable and by the bracket to the cable. The internal forces lie within the cable and are the same magnitude as the external forces but act in the opposite direction, as shown in the free-body diagram in Figure 4.2(b). Since the trap-door is stationary the internal forces balance the external forces and the system is in equilibrium.

### 4.2.4 Distributed forces

All forces fall into one of two categories. **Concentrated forces**, which have been considered so far, are forces that can be assumed to be applied at

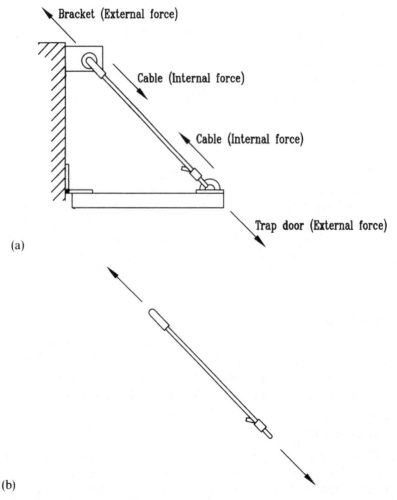

(a)

(b)

**Fig. 4.2** (a) Space diagram of a suspended trap-door; (b) free-body diagram depicting suspension cable.

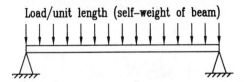

**Fig. 4.3** Self-weight of a beam.

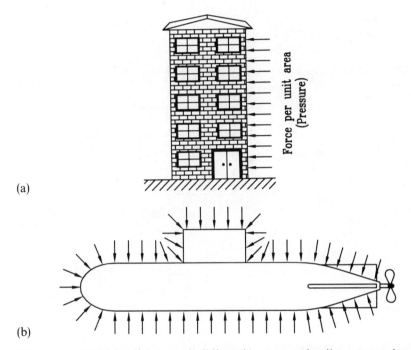

(a)

(b)

**Fig. 4.4** (a) Wind loading on a building; (b) pressure loading on a submarine hull.

a point or on a relatively small area. **Distributed forces** are those that are spread over a length or area. The self-weight of a beam is a typical example of a distributed force, as shown in Figure 4.3.

Wind loading on buildings or the loading of water pressure on the hull of a submarine are both examples of distributed loading of forces over an area, as shown in Figures 4.4(a) and (b).

### 4.2.5 Internally distributed forces

If a tank of water is supported by a vertical rod suspended from above, the tank of water applies a downward force on the rod, as shown in Figure 4.5. The force is distributed over the whole of the cross-section of the rod and must, therefore, be described in terms of an applied force

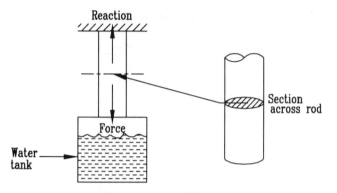

**Fig. 4.5** Water tank supported by a vertical rod in tension.

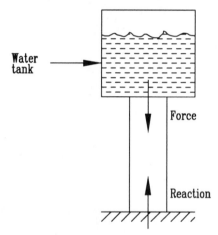

**Fig. 4.6** Water tank supported by a vertical rod in compression.

distributed over an area. This is termed **stress**, designated $\sigma$ (sigma), and takes the general form

$$\text{stress} = \sigma = \frac{\text{force}}{\text{area}}$$

and is defined in units of N/m² (Pa).

If the force is 'pulling' the rod apart, then the stress is termed a **tensile stress**. Alternatively, **compressive stress** results if the forces are 'pushing' on the rod as shown in Figure 4.6, where the water tank is pushing down on the supporting rod.

Calculated values of stress often involve very large quantities and it is usually more convenient to express stress in terms of kilopascals (kPa), or megapascals (MPa), where

$$1\,\text{kPa} = 1000\,\text{Pa}$$
$$1\,\text{MPa} = 1\,000\,000\,\text{Pa}$$

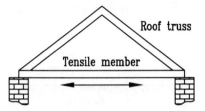

**Fig. 4.7**

## Example 4.1

A tensile member in a roof truss has a cross-sectional area of $150\,\text{mm}^2$ and is subjected to a measured tensile force of $12\,\text{kN}$ (Figure 4.7). Calculate the stress in the member in Pa.

*Solution*

The stress can be found using the given values of force and area:

$$\sigma = \frac{\text{force}}{\text{area}} = \frac{12 \times 10^3}{150 \times 10^{-6}} = 80 \times 10^6\,\text{Pa}$$

$$= 80\,\text{MPa}$$

### 4.2.6 Friction forces

Friction forces are discussed in detail in Chapter 6. However, since friction forces often form all or part of a force system, it is necessary to consider them in the context of 'statics'.

When two surfaces are in contact and are sliding against each other the 'phenomenon of friction' tries to prevent motion. For example, a parent pulls a child sitting on a sledge. In order to move the sledge from rest the parent has to apply a force to the rope to overcome the resistance due to friction. Once the friction is overcome and the sledge is moving, any further increase in the force will merely accelerate the sledge. If two children are sitting on the sledge, the parent has to pull twice as hard to move the sledge. This shows that the greater the load, the greater the force needed to overcome friction. A basic frictional system can be idealized as in Figure 4.8, where $N = $ force applied due to the weight of the block and is always at right angles to the friction plane and $F_r = $ friction force (resisting force due to friction), as explained later in section 6.2.

From these observations the following equation can be deduced and is developed in more detail in Chapter 6:

$$F_r = \mu N \tag{6.2}$$

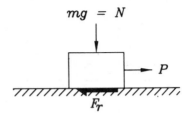

**Fig. 4.8** Basic friction system.

## 4.3 VECTOR REPRESENTATION OF A FORCE (TWO-DIMENSIONAL)

A force has the following characteristics:

1. magnitude
2. direction or line of action
3. point of application.

In Figure 4.9(a) the force in the rope, which suspends the block, is exerting a force $F$ in a vertical direction.

This force $F$ can be represented by a vector since a vector requires both magnitude and direction for it to be valid.

Suppose the block weighs 50 N. The force in the rope will be 50 N at all points along its length and can be depicted by a line whose scale length represents the 50 N. An arrowhead shows the direction in which the force is applied, as shown in Figure 4.9(b).

Since forces always act in pairs, a force and its reaction, there is a force $W$ acting downwards, of 50 N, due to the weight of the block. It is represented by a vector pointing downwards. There is also a reactive force $F$, supporting the weight and directed upwards. The representing vector should also follow that direction. It is said to show the sense of the force. The direction of each force may be distinguished as shown in Figure 4.10.

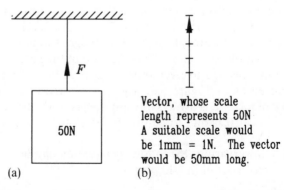

**Fig. 4.9** (a) Weight suspended by a rope; (b) vector representing rope tension.

Figure 4.10(a) represents the weight of the block and Figure 4.10(b) the force in the rope.

Although the forces are equal in magnitude, their directions differ and this can be shown in the notation: force $W_{ab}$ has the vector running from a to b and force $F_{ba}$ has the force running from b to a.

Forces and force systems may be represented and analysed using this method.

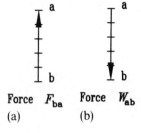

Fig. 4.10

## 4.4 EQUILIBRIUM

The suspended block in Figure 4.9(a) is hanging on the end of a rope and is stationary. The force applied by the weight of the block on the rope is precisely reacted to by the opposing force within the rope. Since the applied force and the reaction force are exactly equal the system can be said to be in equilibrium (in balance). This state can be confirmed since the block does not move up or down. A slight imbalance in any of the forces and the block would rise or fall. The state of equilibrium occurs naturally and has many examples. For example, a chair pushes upwards with the same force as a seated person pushes down.

Thus far equilibrium has only been treated in terms of static forces. However, it should be noted that equilibrium can also exist in moving objects.

A locomotive, pulling carriages and travelling at a constant speed, is in equilibrium since the force developed by the locomotive is exactly equal to the resisting force applied by the carriages. Carriage resistance is due to frictional and air resistances. If the locomotive were to increase its power output, thus increasing the force applied to the rails, between an imbalance the applied and resisting forces would exist and the whole train would increase its velocity.

Attention in this chapter will be confined to equilibrium in static systems.

### 4.4.1 Triangle of forces

It was previously suggested that a force cannot exist alone; it must be paired with a reactive force to be in equilibrium. While this is true, it should be noted that the reaction may be made up of a number of forces. The simplest form of this arrangement is where there are two forces combining to form the equivalent of a single reaction.

Figure 4.11(a) shows a weight of 100 N supported by two cables. The forces in the cables may be analysed by means of the free-body diagram in Figure 4.11(b). Since it can be assumed that the whole system is in equilibrium it can be seen that the force vectors $P$ and $F$ will combine to create an equal and opposite vertical reaction to the force vector $W$.

The forces may be combined in a **triangle of forces** as shown in Figure 4.11(c), where the unknown magnitudes may be graphically determined. By scaling the diagram the magnitudes of the forces in the cables are

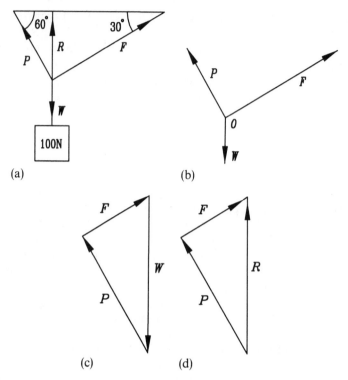

(a)

(b)

(c)

(d)

**Fig. 4.11** A 100 N block supported by two cables.

found to be

$$F = 50 \text{ N}$$
$$P = 87 \text{ N}$$

Although the graphical representation shown in Figure 4.11(c) is a useful means of visualizing the situation, the magnitude of the forces can be found quite easily using trigonometry.

### 4.4.2 Resultant and equilibriant

The forces $P$ and $F$, which provide the reaction, may also be combined into a triangle of forces, Figure 4.11(d), which shows the vertical reaction only. Though the two triangle of forces systems appear to be identical it is important to note that Figure 4.11(c) includes the 'action' force $W$, which is termed the **equilibriant** since it balances the system. Figure 4.11(d) includes the 'reaction' $R$, which is termed the **resultant**. Both the equilibriant and the resultant possess the same magnitude but fulfil different functions and, therefore, lie in opposite directions.

Vector diagrams may be used to analyse multiple force systems, but the basic principles always apply:

• If a number of forces are in equilibrium, the vector diagram must close.

• For equilibrium the vector sum of the forces must equal zero:

$$\Sigma \mathbf{F} = 0 \tag{4.1}$$

or

$$\Sigma \mathbf{F} + \mathbf{P} + \mathbf{W} = 0 \tag{4.2}$$

*Note*: the bold type for the force $\mathbf{F}$ denotes a vector quantity, i.e. it has magnitude and direction.

### 4.4.3 Concurrency

The system shown in Figure 4.11(a) is in equilibrium under the action of three forces in the same plane. When this occurs the lines of action pass through the single point O as shown in Figure 4.11(b). In this case the forces can be said to be **concurrent**. It should be noted that concurrency is a special case and not all force systems may be concurrent.

There are two rules which ensure that a force system is concurrent:

1. The force diagram will close for any set of forces in equilibrium.
2. The criterion of zero moment, equilibrium of moments, is also required.

The rules for concurrent forces ensure that for a three-force system in equilibrium, the vector diagram will close. This, then, provides ground rules which may be used in calculation.

### Example 4.2

A barge is pulled by two tug-boats A and B. The resistance offered by the barge is 30 kN (Figure 4.12). Find the force in each of the tow lines:

(a) graphically
(b) using trigonometry.

*Solution*

(a) By graphical means: both the magnitude and direction of the force on the barge are known so it is possible to draw this vector first to a known scale. Only the directions of $F_A$ and $F_B$ are known so it is possible to draw them on the diagram with arbitrary lengths, thus creating a triangle as shown in Figure 4.12(c). It is then possible to scale $F_A$ and $F_B$ by measuring with a rule:

force $F_A = 20.5\,\text{kN}$
$\qquad F_B = 16.0\,\text{kN}$

(b) Using trigonometry (sine rule):

$$\frac{a}{\sin A} = \frac{b}{\sin B} = \frac{c}{\sin C}$$

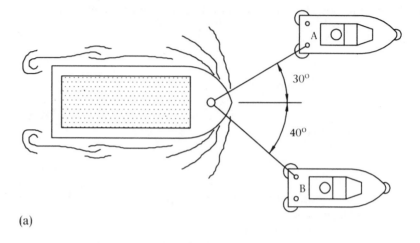

(a)

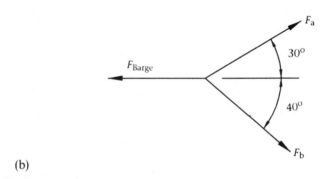

(b)

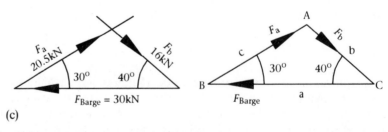

(c)

**Fig. 4.12** (a) Space diagram; (b) free body diagram; (c) vector diagrams.

Find angle $A$

$180° = 30° + 40° + A$

$A = 110°$

Find force $F_A$ ($c$ on Figure 4.12(d))

$$\frac{a}{\sin A} = \frac{c}{\sin C} \quad \text{or} \quad c = \frac{a \sin C}{\sin A}$$

$$= \frac{30 \sin 40°}{\sin 110°} = 20.5 \, \text{kN}$$

Find force $F_B$ ($b$ on Figure 4.12(d))

$$\frac{a}{\sin A} = \frac{b}{\sin B} \quad \text{or} \quad b = \frac{a \sin B}{\sin A}$$

$$= \frac{30 \sin 30°}{\sin 110°} = 16.0 \, \text{kN}$$

Therefore

force $F_A = 20.5 \, \text{kN}$

$F_B = 16.0 \, \text{kN}$

### 4.4.4  Parallelogram of forces

The magnitude and direction of the forces $P$ and $F$ in the cables in Figure 4.11(a) can also be analysed by constructing a parallelogram as shown in Figure 4.13.

The vectors representing the forces $P$ and $F$ are first drawn to scale as they appear on the force system. Parallel lines are then drawn to complete the parallelogram. The final stage is to draw the diagonal $R$. This is not a third independent force, but a single force which replaces $P$ and $F$ and is noted as such by being drawn using two arrowheads.

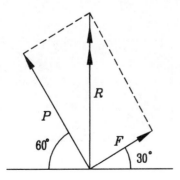

**Fig. 4.13** Parallelogram of forces.

Comparison with the triangle of forces approach shows that Figures 4.11(d) and 4.13 are similar. In fact both methods reveal exactly the same answers. The method used to solve such problems is merely a matter of preference.

### 4.4.5 Resolution of forces

Thus far the discussion has concentrated on combining two forces into a single replacement force, a resultant. It is often useful to split a single force into two components. These are usually at right angles and their vector sum is equal to the original force vector in both magnitude and direction. They are then referred to as **rectangular components**.

The vector representing a force $R$ in Figure 4.14(a) may be resolved into rectangular components $X$ and $Y$, as shown in Figure 4.14(b). Resolution and referral to right-angle axes may also be achieved as shown in Figure 4.14(c).

### Example 4.3

An eye-bolt is loaded by two forces, $F_1 = 30\,\text{kN}$ and $F_2 = 15\,\text{kN}$ (Figure 4.15(a)). Find the magnitude and direction of the resultant force $R$.

To solve this problem analytically it is convenient to consider the applied forces and the resultant as part of a parallelogram of forces diagram (Figure 4.15(b)).

*Solution*

Since sides $a$, $b$ and angle $C$ (Figure 4.16) are known, the resultant can be found by using the cosine rule:

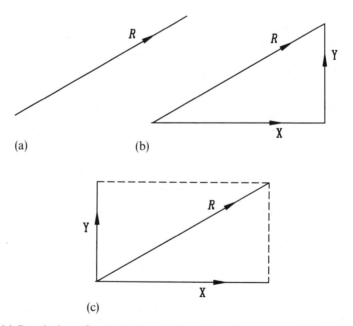

**Fig. 4.14** Resolution of a single force into rectangular components.

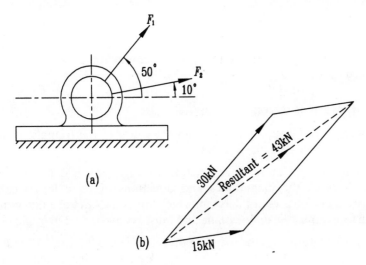

(a)

(b)

**Fig. 4.15** (a) Eye-bolt; (b) triangle of forces.

$$c^2 = a^2 + b^2 - 2ab \cos C$$
$$= 30^2 + 15^2 - 2 \times 30 \times 15 \cos(180° - 50° + 10°)$$
$$c = 42.6 \, \text{kN}$$

To find the angle of the resultant to the horizontal, i.e. its direction, it is possible to use the sine rule:

$$\frac{a}{\sin A} = \frac{c}{\sin C}$$

$$\frac{30}{\sin A} = \frac{42.6}{\sin 140°}$$

and

$$\sin A = \frac{30}{42.6} \times \sin 140°$$

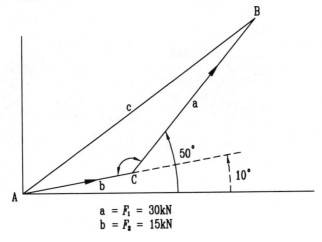

$$a = F_1 = 30\text{kN}$$
$$b = F_2 = 15\text{kN}$$

**Fig. 4.16**

so that

    angle $A = \sin^{-1} 0.4527 = 26.9°$

Therefore, the angle of the resultant to the horizontal is

    $26.9° + 10° = 36.9°$

### 4.4.6 Polygon of forces

The **polygon of forces** approach is an extension of the triangle of forces theorem and can be defined as follows:

> When four or more forces, acting through a single point, are in equilibrium, the magnitudes and directions of these forces can be represented on a vector diagram which forms the sides of a polygon. All the forces must lie in one plane and must be considered in cyclic order.

Figure 4.17 illustrates the use of a polygon of forces. Figure 4.17(a) shows the applied forces $A$, $B$, $C$ and $D$. As they are shown, the system is not in equilibrium as there must be a resultant force acting towards the right.

For the system to be in equilibrium there must be another force, $E$ acting towards the left in order to balance the combined action of the four given forces. The magnitude and direction of force $E$ can be found by drawing a polygon of forces as shown in Figure 4.17(b).

Drawing the vectors, representing forces $A$, $B$, $C$ and $D$, to scale as shown, there is a gap between the two force vectors $A$ and $D$. The vector representing force $E$ can be drawn in so that the polygon is closed, and this gives the magnitude and direction of the force necessary to achieve equilibrium.

Clearly the value of the force $E$ can be found graphically, simply by drawing the polygon of forces, using a ruler and a protractor. The accuracy of the values of the force and its angle with respect to the datum, in this case force $A$, depends upon how accurately the diagram is drawn.

There is an alternative analytical approach to this type of problem, as described in the next section.

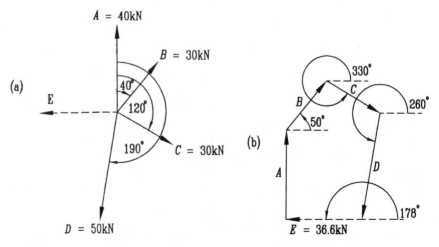

**Fig. 4.17** Polygon of forces.

## 4.5 ANALYSIS OF FORCES

As described earlier, any combination of forces that act through a single point can be resolved into a single resultant force.

Consider the polygon of forces shown in Figure 4.18. The system consists of three applied forces, $A$, $B$ and $C$. These can be resolved into a resultant $R$ as shown. The vertical components of all the applied forces must be equal to the vertical component of the resultant. This can be stated as

$$R_V = A_V + B_V + C_V$$

It can be shown that this is true for any number of forces, so that

$$R_V = \sum F_V \tag{4.3}$$

Similarly it can be shown that the horizontal component of the resultant is equal to the sum of the horizontal components of all the forces:

$$R_H = \sum F_H \tag{4.4}$$

Equations (4.3) and (4.4) can be used as the basis of an analytical method of evaluating the magnitude and direction of the resultant. In order to do so it is usual to find the vertical and horizontal components of each force.

Figure 4.19 shows the vertical and horizontal components of a force $F$ inclined at an angle $\theta$ to the horizontal axis. The two components can be found by:

$$F_V = F \sin \theta \tag{4.5}$$

and

$$F_H = F \cos \theta \tag{4.6}$$

The value of the components and their respective signs, whether positive or negative, can be found directly by measuring the angle from the horizontal datum shown in Figure 4.19.

The most straightforward way of applying this approach is to tabulate the components as shown in Table 4.1. This gives values of $R_V$ and $R_H$ from which the resultant can be found as illustrated in Example 4.4. Alternatively it is possible to show that a number of forces are in equilibrium if the values of $R_V$ and $R_H$ are zero.

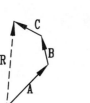
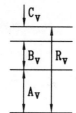
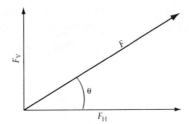

**Fig. 4.18** Vertical components of forces.

**Fig. 4.19** Vertical and horizontal components of a force.

**Table 4.1**

| F | $F_V$ | $F_H$ |
|---|---|---|
| A | $A_V$ | $A_H$ |
| B | $B_V$ | $B_H$ |
| ,, | ,, | ,, |
| ,, | ,, | ,, |
| R | $R_V$ | $R_H$ |

### Example 4.4

Evaluate the magnitude and direction of the resultant of the forces defined in Figure 4.20.

*Solution*

The vertical and horizontal components of forces *A*, *B*, *C* and *D* can be found using equations (4.5) and (4.6), as follows:

| F | $\theta$ | $F_V$ | $F_H$ |
|---|---|---|---|
| A | 40 | 90° | 40 | 0 |
| B | 30 | 50° | 22.98 | 19.28 |
| C | 30 | 330° | $-15.0$ | 25.98 |
| D | 50 | 260° | $-49.24$ | $-8.68$ |
| R | | | $-1.26$ | 36.58 |

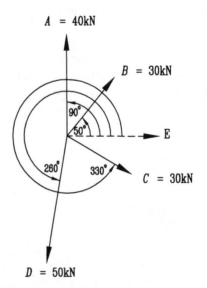

Fig. 4.20

Using trigonometry:

$$\tan \theta = \frac{F_V}{F_H} = \frac{-1.26}{36.58} = -0.0345$$

Therefore

$$\theta = \tan^{-1}(-0.0345) = -2° \quad \text{i.e. } 358°$$

and

$$F = \frac{F_V}{\sin \theta} = \frac{-1.26}{\sin 358°} = 36.1 \, \text{kN}$$

## Example 4.5

The members of a truss are connected to the gusset plate, as shown in Figure 4.21. Determine the force $T$ and its angle to the horizontal, to put the truss into equilibrium.

*Solution*

The vertical and horizontal components of forces $A$, $B$ and $C$ can be found using equations (4.5) and (4.6), as follows:

|   | $F$ | $\theta°$ | $F_V$ | $F_H$ |
|---|---|---|---|---|
| $A$ | 9 | 15° | 2.33 | 8.69 |
| $B$ | 12 | 85° | 11.95 | 1.05 |
| $C$ | 6 | 150° | 3.0 | −5.2 |
| $T$ | ? | ? | −17.28 | −4.54 |
| $R$ | | | 0 | 0 |

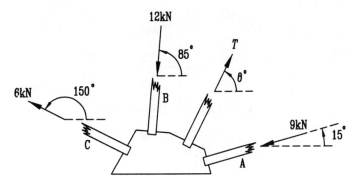

Fig. 4.21

so that $T = \sqrt{(-17.28)^2 + (-4.54)^2} = 17.87 \, \text{kN}$ and the angle of $T$ to the horizontal is:

$$\theta = \tan^{-1}\left(\frac{-17.28}{-4.54}\right) = 75.28°$$

## 4.6 MOMENTS

### 4.6.1 Turning effects of a force

If a force is applied to a component, such as the disc shown in Figure 4.22, and it is not directed through the axis, then the disc will tend to turn. Furthermore, the greater the distance of the force from the axis, the greater the turning effect. This turning effect is termed the **moment** and is designated $M$. The magnitude of the moment is, therefore

moment $= M =$ force $\times$ radius

or

$$M = F \times r \tag{4.7}$$

The units are newton metres (N m).

*Note:* in all cases the force is applied tangentially (at 90°) to the line which represents the radius.

The imaginary line which joins the point of application of the force to the axis is called the **moment arm** and is illustrated in Figure 4.23 as a crank handle.

There are many examples of turning moments, such as a door handle, the turning of a key in a lock, turning a steering wheel, a screwdriver turning a screw and a spanner turning a nut.

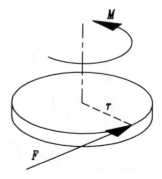

**Fig. 4.22** The turning effect of a force applied at a radius.

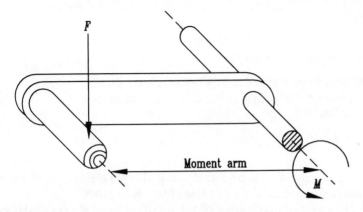

**Fig. 4.23** Crank handle demonstrating a moment arm.

### 4.6.2 Couple

A couple is a special type of moment where there is no point at which the axis can be fixed. The couple can be considered to be derived from two equal and opposite forces separated by a distance $r$, as shown in Figure 4.24.

This may be thought of as a piece of plywood floating in water and being turned using a finger of each hand. The position of the axis of rotation would vary as the board turned in the water.

A couple can, therefore, be defined as

couple = force × distance between forces

or

$$C = F \times r \tag{4.8}$$

The units are newton metres, as before.

### 4.6.3 Equivalent force/moment systems

It is often convenient to combine multiple forces into a single force and a single moment acting on a component as outlined in Example 4.6.

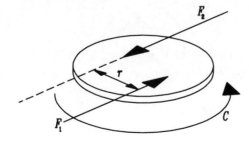

**Fig. 4.24** Couple applied to a disc in free space.

### Example 4.6

The two forces $F_1$ and $F_2$ shown in Figure 4.25 have values of 110 and 240 N respectively. Determine:

(a) the combined force $R$ acting towards the axis;
(b) the moment $M$ acting at the end of the moment arm whose length is $r = 0.5$ m.

*Solution*

(a) The resultant $R$ can be found using the parallelogram of forces, as shown. The value of the resultant force $R$ is 310 N.
   The point of application of the resultant force $R$ acts at the axis of rotation as shown in the force/moment diagram (Figure 4.25(c)).
(b) In order to find the total moment, the moments applied by the forces must be algebraically summed.
   To derive the direction assume one of the directions is positive. In this case the larger force suggests that the positive direction should be anticlockwise. Any direction may be designated positive, however this direction should be consistent throughout the calculation.
   The two forces $F_1$ and $F_2$ are trying to turn the disc in opposite directions. The turning moment is, therefore, the difference between the

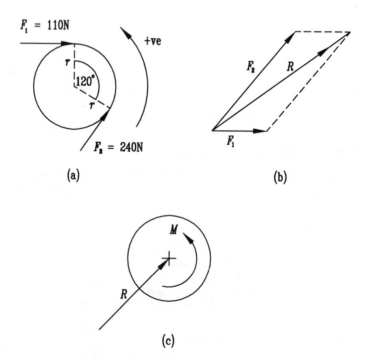

**Fig. 4.25** (a) Space diagram; (b) vector diagram; (c) force/moment diagram.

two forces multiplied by the moment arm $r$, i.e.

moment $=$ force $\times$ moment arm

$$M = (F_2 - F_1) \times r$$
$$= (240 - 110) \times 0.5 = 65\,\text{N m}$$

Alternatively, it can be said that the algebraic sum of the moments will give the total moment and its direction:

total moment $=$ anticlockwise moment $-$ clockwise moment

or

$$M = (F_2 \times r) - (F_1 \times r)$$
$$= (240 \times 0.5) - (110 \times 0.5)$$
$$= 120 - 55 = 65\,\text{N m}$$

*Note:* the two forces $F_1$ and $F_2$ have now been combined into a representative single force and single moment.

Generally, the vector addition of the forces gives the combined force and direction, i.e.

$$\mathbf{R} = \mathbf{F}_1 + \mathbf{F}_2 \tag{4.9}$$

or

$$\mathbf{R} = \sum \text{force vectors} \tag{4.10}$$

*Remember:* a bold variable indicates a vector quantity.

For any combination of forces and radii the algebraic addition of all the force $\times$ radius products gives the total moment $M$, i.e.

$$M = (F_1 \times r_1) + (F_2 \times r_2) + (F_3 \times r_3) + \ldots \tag{4.11}$$

or

$$M = \sum (F \times r) \tag{4.12}$$

### 4.6.4 Levers

A lever can be described as a bar which is supported at a single point, called a fulcrum. The fulcrum may take the form of say a set of bearings or merely a knife edge. The lever shown in Figure 4.26 is supported by a knife edge and when there are no applied forces the lever will balance.

If a force $F_1$ is applied to the left-hand side (LHS) at a distance $a$ from the fulcrum, an anticlockwise turning moment will be applied to the lever. In order to restore equilibrium (balance) a force $F_2$ is applied to the RHS but at a distance $b$ from the fulcrum.

Since the distance $b$ is smaller than $a$, it follows that the force $F_2$ must be larger than $F_1$ in order to provide the same magnitude of moment.

For the lever to be in equilibrium the moments on each side of the fulcrum must be equal and can be described thus:

left-hand side $=$ right-hand side

anticlockwise moments $=$ clockwise moments

$$\circlearrowleft = \circlearrowright$$
$$F_1 \times a = F_2 \times b \tag{4.13}$$

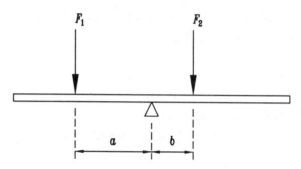

**Fig. 4.26** Lever supported by a knife edge.

Generally this approach may be used to find a single unknown value as described in Example 4.7.

**Example 4.7**

A uniform lever, pivoted at its mid point B, has a mass of 6 kg suspended from point C which is 100 mm to the right of the fulcrum (Figure 4.27). Find the balancing mass to be hung from point A.

*Solution*

In order to determine the balancing mass, moments must be taken about the fulcrum.

$$\text{force at } C = 6 \times 9.81 = 58.9\,\text{N}$$

Taking moments about the fulcrum:

$$\text{anticlockwise moments} = \text{clockwise moments}$$

$$\circlearrowleft = \circlearrowright$$

$$\text{force at A} \times 0.8 = 58.9 \times 0.1$$

$$\text{force at A} = \frac{58.9 \times 0.1}{0.8} = 7.36\,\text{N}$$

$$\text{equivalent mass at A} = \frac{7.36}{9.81} = 0.75\,\text{kg}$$

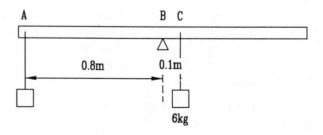

**Fig. 4.27**

It should be noted that the mass of 0.75 kg required to balance the system is significantly less than the original mass of 6 kg. The reduction in mass is, however, in direct proportion to the change in length of the moment arm from 100 to 800 mm.

### 4.6.5 Equilibriant and resultant of parallel forces

A number of forces acting downwards on a lever may be counteracted at a single point. This **equilibriant**, reactive force is equal and opposite to the forces acting downwards. If, however, all the forces acting downwards are combined into a single force, this force would be called the resultant.

The lever shown in Figure 4.28 has three parallel forces acting downwards of magnitude 40, 60 and 30 kN. In order to find the equilibriant it is necessary to find the force acting upwards at A and its position $z$ from the left-hand end.

The three forces, when combined, will give a resultant force of

$$R = 60 + 40 + 30 = 130\,\text{kN}$$

Since the equilibriant is equal and opposite to the resultant of 130 kN, this is also the value of the equilibriant acting upwards at the fulcrum.

In order to determine the position of the fulcrum $z$, moments must be taken. The 'hinge' or point about which the moments may be taken can be placed at a convenient point. In this case it is the point of application of the 40 kN force.

anticlockwise moments = clockwise moments

$$\circlearrowleft = \circlearrowright$$

$$A \times z = (60 \times 1.5) + (30 \times 2.3)$$

$$130 \times z = (60 \times 1.5) + (30 \times 2.3)$$

$$z = \frac{(60 \times 1.5) + (30 \times 2.3)}{130}$$

$$= \frac{159}{130} = 1.22\,\text{m}$$

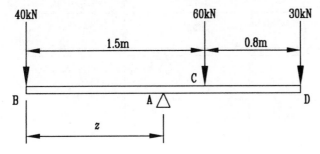

**Fig. 4.28** Lever under the influence of three vertical forces.

For complete equilibrium the fulcrum A needs to be applied at a distance of 1.22 m to the right of point B. This means that the same force acts upwards as acts downwards and the beam is in vertical equilibrium. This also means that the beam is balanced about the fulcrum, that is, it is in rotational equilibrium.

### 4.6.6 The general principle of moments

The example in section 4.6.5 was solved by taking moments about the point of application of one of the forces; however, the point around which moments may be taken can be any convenient point.

Consider the example in Figure 4.29. This is exactly the same lever as shown in Figure 4.28 except that an arbitrary hinge point has been taken.

It is necessary to determine the position of the fulcrum A.

Find the resultant by first summing the vertical forces

$$R = 40 + 60 + 30 = 130\,\text{kN}$$

Find the distance $z$ by taking moments about the hinge point as follows:

$$\text{anticlockwise moments} = \text{clockwise moments}$$

$$\circlearrowleft = \circlearrowright$$

$$(B(p - r)) + (A \times z) = (C \times r) + (D(q + r))$$

$$(40(1.5 - 1.0)) + 130z = (60 \times 1.0) + 30(0.8 + 1.0)$$

$$20 + 130z = 114$$

$$z = \frac{114 - 20}{130} = 0.72\,\text{m}$$

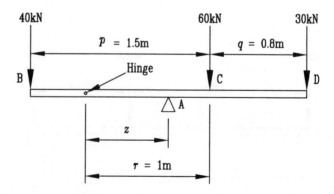

**Fig. 4.29** Lever showing arbitrary hinge point.

The position of the fulcrum should be 0.72 m from the hinge point.

Comparison with the position of the fulcrum in the example in section 4.6.5 will show that the fulcrum in the above example is applied in exactly the same position on the beam.

### 4.6.7  General note on the positioning of the hinge point

The example in section 4.6.5 placed the hinge point at a point of application of one of the forces. Since the force acts directly on the hinge it cannot produce a moment and is, therefore, eliminated from the calculations.

Though the example in section 4.6.6 is exactly the same problem, the hinge point is in mid-span. The previously eliminated force of 40 kN now induces a moment and must be included in the moment equation.

The lesson to be learned here is that the analysis will always be simplified if the hinge is placed at the point of application of a force.

The **general principle of moments** may be defined as follows:

> When body is in equilibrium and under the influence of several forces, the total clockwise moment about any hinge point must equal the total anticlockwise moment.

### 4.6.8  Support reactions in a horizontal beam

The consideration of levers is useful in order to analyse the application of moments. Often, in practice, it is necessary to provide two support points for a uniform length. This then is not a lever but a **beam**. A beam is a uniform length of solid material, either wood or metal, which may span a roof or is a structural member in a bridge. Whatever the function the following may be assumed:

- The beam may support multiple forces acting downwards.
- The beam is supported in two places, usually at each end, but this may not always be the case.
- The support reactions support the weight of the beam and all the applied forces on it.
- The beam is stationary (static).

Since the beam is stationary it is in equilibrium and the principle of moments may be employed.

### Example 4.8

A uniform beam is 4 m long and has applied to it two loads as shown in Figure 4.30. Determine:

(a) the support reaction $R_2$
(b) the support reaction $R_1$.

Assume the mass of the beam to be negligible.

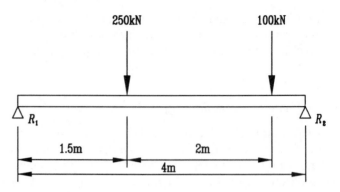

**Fig. 4.30** Uniform beam.

*Solution*

(a) To find the reaction $R_2$ take moments about the point of application of $R_1$. This will eliminate the unknown $R_1$ from this stage of the calculations.

$$\circlearrowleft = \circlearrowright$$

$$R_2 \times 4 = (250 \times 1.5) + (100 \times 3.5)$$

$$R_2 = \frac{(250 \times 1.5) + (100 \times 3.5)}{4}$$

$$= \frac{725}{4} = 181.25 \, \text{kN}$$

(b) To find the reaction $R_1$ there are two possible approaches:
   (i) equilibrium of forces
   (ii) take moments about the point of application of $R_2$.

**Equilibrium of forces approach**
For equilibrium, all the forces acting upwards must equal all the forces acting downwards:

forces acting up = forces acting down

$$R_1 + R_2 = 250 + 100$$

$$R_1 + 181.25 = 350$$

$$R_1 = 350 - 181.25 = 168.75 \, \text{kN}$$

**Taking moments about the point of application of $R_2$**

$$R_1 \times 4 = (250 \times 2.5) + (100 \times 0.5)$$

$$R_1 = \frac{675}{4} = 168.75 \, \text{kN}$$

*Note:* both methods use equilibrium as the basis for the analysis.

## 4.7  CONTACT FORCES

When two surfaces touch there are normally internal forces which arise from the contact. In determining a full set of forces applied to a body it is often necessary to include the contact forces along with the externally applied forces.

Contact forces occur in pairs and are equal and opposite forces exerted on the contacting bodies. The type of contact force, its magnitude and direction, depends upon the type of contact between the adjoining bodies.

The identification of the contact force type is an important part of determining the true picture of the forces acting on a body.

Figure 4.31 shows the most common forms of contact force.

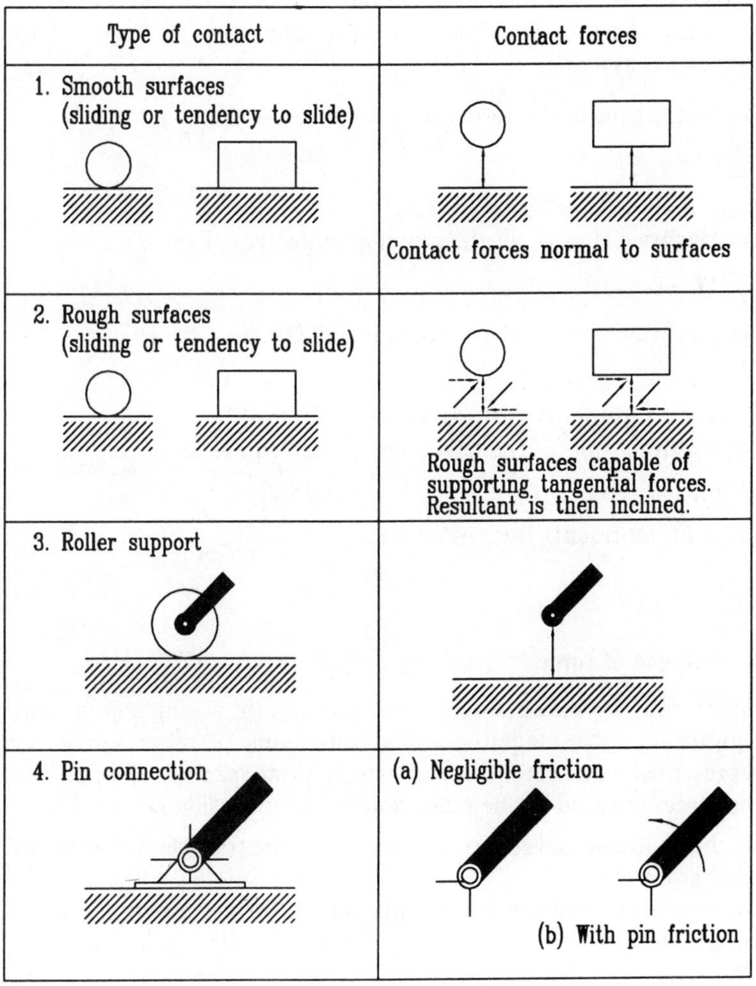

| Type of contact | Contact forces |
| --- | --- |
| 1. Smooth surfaces (sliding or tendency to slide) | Contact forces normal to surfaces |
| 2. Rough surfaces (sliding or tendency to slide) | Rough surfaces capable of supporting tangential forces. Resultant is then inclined. |
| 3. Roller support | |
| 4. Pin connection | (a) Negligible friction  (b) With pin friction |

**Fig. 4.31** Types of contact force.

## 4.8 EQUILIBRIUM SYNOPSIS

Equilibrium was introduced and explained in section 4.4. It is essential to understand the importance of equilibrium since it is the basis for all the calculations performed in this chapter. Indeed it is also at the heart of analysis of many other branches of science, such as thermodynamics, fluid dynamics, kinematics, stress, electronics, chemistry, etc.

The general assumption made in this chapter is that if a system is in equilibrium it must be static. However, there are many situations where the system does not have to be stationary for equilibrium to apply, but the motion must be uniform

Within the scope of this chapter the use of the principles of equilibrium allows the balancing of action forces and moments with reaction forces and moments. For this to be true, force systems have to be considered as static. If these systems were unbalanced they would move, that is, they would be 'dynamic systems' and other, appropriate theory would be required.

Generally, the following 'rules of equilibrium' may be applied to any force/moment system:

- The vector sum of the forces must equal zero, i.e.

$$\sum \mathbf{F} = 0$$

where $\mathbf{F}$ represents vector forces.
- The algebraic sum of all the moments must equal zero, i.e.

$$\sum M = 0$$

- The algebraic sum of the vertical forces equals zero, i.e.

$$\sum F_V = 0$$

where $F_V$ represents vertical forces.
- The algebraic sum of the horizontal forces equals zero, i.e.

$$\sum F_H = 0$$

where $F_H$ represents horizontal forces.

## 4.9 SUMMARY

### 4.9.1 Polygon of forces

When four or more forces, acting through a single point, are in equilibrium, the magnitudes and directions of these forces can be represented on a vector diagram which forms the sides of a polygon. All the forces must lie in one plane and must be considered in cyclic order.

*Note:* this rule also covers, in a general way, the triangle of forces and the parallelogram of forces.

For any vector diagram, for equilibrium, the vector diagram must close.

For equilibrium the vector sum of the forces must equal zero, or

$$\sum \mathbf{F} = 0 \tag{4.1}$$

*Note:* the bold type denotes a vector quantity.

**Table 4.2**

| $F$ | $F_V$ | $F_H$ |
|---|---|---|
| $A$ | $A_V$ | $A_H$ |
| $B$ | $B_V$ | $B_H$ |
| ” | ” | ” |
| ” | ” | ” |
| $R$ | $R_V$ | $R_H$ |

Each force may be analysed as a vertical component and a horizontal component and may be summed for each force, as shown in Table 4.2, where

$$F_V = F \sin \theta \tag{4.5}$$

and

$$F_H = F \cos \theta \tag{4.6}$$

### 4.9.2 The general principle of moments

When a body is in equilibrium and under the influence of several forces, the total clockwise moment about any axis must equal the total anticlockwise moment.

*Moments*

moment $= M =$ force $\times$ radius

$$M = F \times r \tag{4.7}$$

couple $= C =$ force $\times$ distance between forces

$$C = F \times r \tag{4.8}$$

*Levers and beams*

anticlockwise moment $=$ clockwise moment

$$\circlearrowleft = \circlearrowright$$

force $\times$ moment arm $=$ force $\times$ moment arm

Generally,

$$F_1 \times a = F_2 \times b \tag{4.13}$$

where $a$ and $b$ are moment arms.

*Equilibrium*

• The vector sum of the forces must equal zero, i.e.

$$\sum \mathbf{F} = 0$$

where $\mathbf{F}$ represents vector forces.

- The algebraic sum of all the moments must equal zero, i.e.

$$\sum M = 0$$

- The algebraic sum of the vertical forces equals zero, i.e.

$$\sum F_V = 0$$

where $F_V$ represents vertical forces.

- The algebraic sum of the horizontal forces equals zero, i.e.

$$\sum F_H = 0$$

where $F_H$ represents horizontal forces.

### 4.10  PROBLEMS

1. Can it be shown that for any concurrent force system, the addition of only one force is necessary to achieve equilibrium?
2. The belt and pulley shown in Figure 4.32 is influenced by belt tensions of 2500 and 600 N. Find the magnitude and direction of the resultant force which acts on the shaft.
3. The plate shown in Figure 4.33, is subjected to two forces $A$ and $B$. If the angle between the two forces is 70°, determine the magnitude of the resultant force.
4. The gusset plate shown in Figure 4.34 is part of a bridge truss. Using a polygon of forces determine the equilibriant.
5. A heavy packing case is to be suspended by three ropes as shown in Figure 4.35. Determine the magnitude and direction of the force $F$ so that the resultant force is directed vertically upwards and has a magnitude of 900 N.
6. A lever AB is 2.5 m long and is supported at end A. A mass of 4 kg is suspended at a point C, 0.9 m from end A. Calculate the vertical force required at B to hold the lever in horizontal equilibrium.

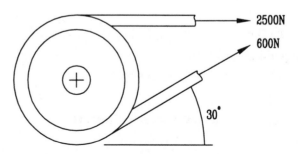

**Fig. 4.32** Belt and pulley system.

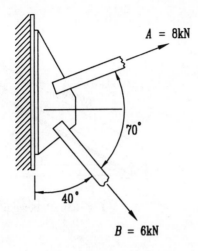

**Fig. 4.33** Gusset plate.

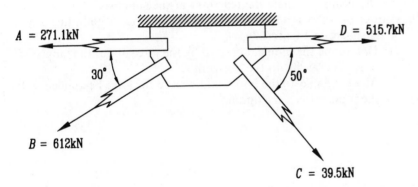

**Fig. 4.34** Gusset plate.

7. A uniform horizontal lever, AE, 7 m long, is supported on a fulcrum at C, which is 3.5 m from A. There are downward forces of 25 and 35 N at A and E respectively and another downward force of 15 N at point B which is 2 m from A. What vertical force must be applied at point D, 1.5 m from E, to maintain the lever in horizontal equilibrium? Will this force act upwards or downwards? Assume the lever is of negligible mass.

8. A horizontal steel beam of uniform cross-section is 5 m long and rests on two supports, one at one end of the beam and the second at 1 m from the other end.
   (a) If the mass of the beam is 120 kg, find the reactions at the supports.
   (b) A 60 kg mass is now suspended midway between the supports. Determine the total reactions at the supports.

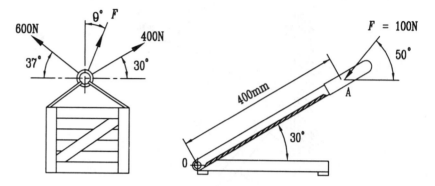

**Fig. 4.35** Packing case.          **Fig. 4.36** Paper guillotine.

9. The mass of a uniform steel girder, 15 m long, is 3 t. It is supported at one end and also at 5 m from the other end.
   (a) Calculate the reactions at the two supports.
   (b) If a mass of 2 t is suspended at the end of the overhanging portion of the beam, calculate the reactions at the supports.
10. A force of 100 N is applied to the handle of a paper guillotine at point A, as shown in Figure 4.36.
   (a) Determine the moment created by the force about the hinge O if it is applied at 50° to the horizontal.
   (b) At what angle, to the horizontal, should the force be applied so that the moment is a maximum?

# Work, energy and power $\boxed{5}$

## 5.1 AIMS

- To introduce the concepts of work, energy and power.
- To define the equations used to analyse work, energy and power.
- To explain how work, energy and power are applied to both linear and angular motion.
- To define the relationship between potential energy and kinetic energy.
- To define efficiency in terms of power.

## 5.2 WORK

A body which possesses energy, perhaps due to its motion or its position, has the capacity to do work. Before energy or power can be discussed it is, therefore, important to understand what is meant by the terms 'to do work' or 'work done'.

Work is done when the point of application of a force is exerted over a distance and can be defined as follows:

> When a force of 1 N is exerted over a distance of 1 m, then a value of 1 N m of work has been expended.

Since the unit of N m is used to define a moment or a torque, the unit of work is termed the joule and designated J.

$$\text{work done} = \text{force} \times \text{distance}$$
$$w = F \times s \quad \text{(J)} \tag{5.1}$$

### Example 5.1

At constant velocity a car is opposed by a resistive force of 150 N. Find the work done in moving the car through a distance of 2 km.

*Solution*

The 'work done' can be found by using equation (5.1):

$$w = F \times s$$

Now $F = 150\,\text{N}$ and $s = 2000\,\text{m}$ so that

$$w = 150 \times 2000 = 300\,000\,\text{J} = 300\,\text{kJ}$$

### 5.2.1 Work represented by a diagram

Figure 5.1 shows a graph whose vertical axis represents force and whose horizontal axis represents distance. The area enclosed represents the total work done. Here, the point of application of a force of 50 N moves through a distance of 12 m, giving a total value of $w$ of 600 J.

Work done may also be represented by the area enclosed under a curve on a graph. Consider a spring which is gradually extended by applying a force of increasing magnitude. If the extension of the spring and the load are plotted on a force/extension graph, as in Figure 5.2, a straight line can be plotted. The area under the graph line represents the work done.

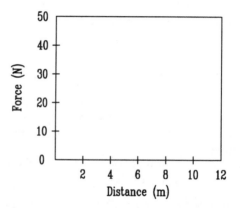

**Fig. 5.1** Work represented by area.

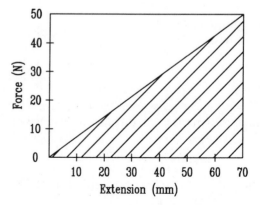

**Fig. 5.2** Force extension diagram for a spring.

**Example 5.2**

Determine the work done when a spring extends 70 mm when it is uniformly loaded from zero up to 50 N.

*Solution*

The 'work done' can be found by drawing a force/extension diagram, as shown in Figure 5.2. Now

$$\text{total force} = F = 50\,\text{N}$$

$$\text{total extension} = 70\,\text{mm} \quad \text{or} \quad 0.07\,\text{m}$$

so that

$$w = \frac{F \times \text{extension}}{2}$$

$$= \frac{50 \times 0.07}{2} = 1.75\,\text{J}$$

### 5.2.2 Work done by a varying load

The 'work done' found in Example 5.2 is the total work done, since this varies with the force as the extension of the spring takes place. Though this type of graph is useful for calculating uniformly varying loads and extensions, it becomes even more effective where forces vary in a non-uniform way. The mid-ordinate rule can be used to calculate the average force and, hence, determine the work done. Equation (5.1) may be modified as follows:

$$w = \text{average force} \times \text{distance}$$

or

$$w = F_{AV} \times s \tag{5.2}$$

**Example 5.3**

A load is hauled with varying tractive effort of $F$ newtons over a distance according to the following table:

| $x$ (m) | 0 | 20 | 50 | 80 | 110 | 130 | 160 | 190 | 200 |
|---|---|---|---|---|---|---|---|---|---|
| $F$ (N) | 1280 | 1270 | 1220 | 1110 | 905 | 800 | 720 | 670 | 660 |

Find:

(a) the average force;
(b) the work done in kJ.

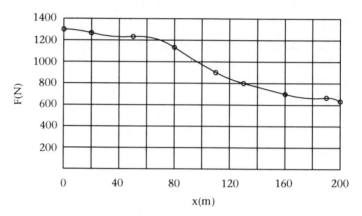

**Fig. 5.3** Plot of force against distance.

*Solution*

(a) The average force can be found by plotting $F$ against $x$, as shown in Figure 5.3, and using the mid-ordinate rule to perform the calculation.

Determine the height of the graph line at the mid-point (mid-ordinate) of each 20 m strip, i.e. take a reading of the force at 10, 30, 50 m, etc., as follows:

| $x$ (m) | 10 | 30 | 50 | 70 | 90 | 110 | 130 | 150 | 170 | 190 |
|---|---|---|---|---|---|---|---|---|---|---|
| $F$ (N) | 1275 | 1260 | 1250 | 1200 | 1050 | 905 | 800 | 740 | 695 | 670 |

This gives 10 readings in all.

$$\text{average force} = F_{AV} = \frac{\Sigma F}{10}$$

$$= \frac{9845}{10} = 984.5\,\text{N}$$

It should be noted that the greater the number of ordinates values, the more accurate the calculation of the average force.

(b) The work done can be calculated using equation (5.2):

$$w = F_{AV} \times s$$
$$= 984.5 \times 200 = 196\,900\,\text{J}$$
$$= 196.9\,\text{kJ}$$

### 5.2.3  Work done by an oblique force

It may happen that the line of action of the force is at an angle to the direction of motion of the body. In this case the force should be resolved into two components, one normal to the motion and one parallel to the motion as shown in Figure 5.4.

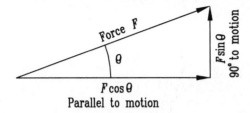

**Fig. 5.4** Resolving an oblique force into two components.

The component $F \sin \theta$, at $90°$ to the motion, does not contribute to the forward motion and can be ignored. The work done from an oblique force is, therefore, developed by $F \cos \theta$ alone and can be inserted into equation (5.1) as follows:

$$w = F \cos \theta \times s \qquad (5.3)$$

**Example 5.4**

A barge is being towed along a canal by a horse at a speed of 5 km/h. The tow rope is inclined at $25°$ to the direction of motion, as shown in Figure 5.5, and the horse causes a steady tension in the rope of 300 N. Find:

(a) the distance travelled in 2 min;
(b) the work done during 2 min of exertion.

*Solution*

(a) The distance travelled in 2 min can be found by using equation (2.1):

$$v = \frac{s}{t}$$

Now

$$v = 5 \,\text{km/h} \quad \text{or} \quad \frac{5 \times 1000}{60 \times 60} = 1.39 \,\text{m/s}$$

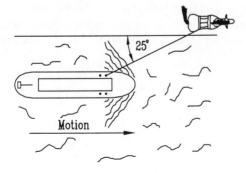

**Fig. 5.5** Horse pulling a barge along a canal.

and

$t = 2\,\text{min or } 120\,\text{s}$, so that

$$1.39 = \frac{s}{120}$$

or

$$s = 120 \times 1.39 = 166.7\,\text{m}$$

(b) The work done can be found by using equation (5.3):

$$w = F \cos \theta \times s$$

Now

$F = 300\,\text{N}, s = 166.7\,\text{m and } \theta = 25°$, so that

$$w = 300 \times \cos 25° \times 166.7 = 45\,320\,\text{J}$$
$$= 45.32\,\text{kJ}$$

### 5.2.4 Work done in rotation

Consider a block of wood being pulled across the floor by a rope. A force is applied to the block when moving through a distance. Work is being done on the system in order to move the block.

The rope is now wrapped around a pulley which applies the same tension in the rope as before, and takes up the length of rope as the block moves across the floor. The pulley system performs the same work as before, but this is done in an angular sense, and may be deduced from equation (5.1):

$$w = F \times s$$

By considering the angular equivalents of force and displacement, namely torque and angular displacement, the work done in rotation may be described as follows:

$$w = T \times \theta \tag{5.4}$$

The force in the rope is applied at the outside radius of the pulley, hence a torque is applied. The linear displacement is now equivalent to the rope being wrapped around the pulley which is the angular displacement. The derivation of equation (5.4) can, therefore, be described as follows:

$$w = \text{force} \times \text{circumference}$$

$$= F \times 2\pi r$$

or

$$w = (F \times r) \times 2\pi$$

and

$$w = T \times 2\pi$$

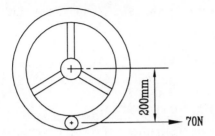

**Fig. 5.6** Hand crank wheel.

but $2\pi$ is 1 rev expressed in rad, so

$$w = T \times \theta$$

Here the derivation uses one complete revolution; however, it is true that any angular value can be inserted in place of $\theta$ provided that the value is in rad.

**Example 5.5**

A pulley system is to be turned by hand crank, as shown in Figure 5.6. The handle is placed at a radius of 200 mm from the centre of rotation and a force of 70 N is applied to the handle.

Determine the work done in one revolution of the crank. Assume that the force is always applied at 90° to the crank arm.

*Solution*

The work done in one revolution can be found by using equation (5.4):

$$w = T \times \theta$$

Now

$$T = F \times r = 70 \times 0.2 = 14 \, \text{N m}$$

and

$$\theta = 1 \, \text{rev} = 2\pi \, \text{rad}$$

so that

$$w = 14 \times 2\pi = 88 \, \text{J}$$

**5.2.5  Work done in an angular sense**

It is still very common to quote angular displacement and angular velocity in terms of revolutions of a system. It is necessary to convert rev, for angular displacement, and rev/min or rev/s, for angular velocity, into units which are part of the SI system. These are rad, for angular displacement, and rad/s for angular velocity.

Since there are $2\pi$ (6.284) rad in every rev, it follows that equation (5.4) can be rewritten as follows:

$$w = T \times \theta$$

$$= T \times 2\pi \quad \text{(J)} \quad \text{for every rev}$$

If $n$ now represents the number of rev, the equation becomes

$$w = 2\pi T n \quad \text{(J)} \quad \text{for } n \text{ rev} \tag{5.5}$$

**Example 5.6**

A winch capstan of 800 mm diameter applies a tension in a rope of 2000 N. If the capstan speed is 5 rev/min, find the work done, in kJ, in 8 min.

*Solution*

The work done in 8 min can be found using equation (5.5):

$$w = 2\pi T n$$

Now

$$T = F \times r = 2000 \times 0.4 = 800 \, \text{Nm}$$

and

$$n = 5 \times 8 = 40 \, \text{rev}$$

so that

$$w = 2\pi \times 800 \times 40 = 201 \, \text{kJ}$$

## 5.3 ENERGY

Energy can be defined as 'that state of a body which gives it the "capacity" to do work'. Various forms of energy can be harnessed to do work, and include: chemical, nuclear, electrical and solar. For example, chemical energy is released when petrol is burned in an internal combustion engine. Compressed air possesses energy due to its high pressure and is used to operate power tools. High-temperature steam is used to turn turbines by virtue of its thermal energy. Though there are many applications of energy in its various forms, it is the application of mechanical energy that is important in the present discussion.

Mechanical energy takes two forms. *Potential energy* is possessed by a body due to its position or state, while *kinetic energy* is possessed by a body due to its velocity. The term 'potential energy' often refers to the energy possessed by a body due to its height above some datum. This can be considered as 'gravitational' potential energy. Potential energy may also be used to describe other systems where energy is stored. For example, a compressed spring possesses potential energy because it has the potential of doing work. When released the spring has the opportunity of performing the work. Work has to be done in order initially to compress the spring. The energy stored in the spring is really elastic potential energy and is often termed 'strain energy'.

Since work has to be done to provide a body with energy and energy has to be expended in order to do work, work and energy are interchangeable quantities and share the same units of the joule (J).

### 5.3.1 Potential energy

Potential energy is concerned with 'state' or 'position' relative to some convenient datum. An example of potential energy of 'state' is that of a pressurized air cylinder, where the pressurized state of the air has the potential of providing energy. An example of potential energy due to position is that of a mass at some height above the ground. Here the mass has potential of doing work by falling. It should be noted that work has to be done to raise the mass in order to give the mass its potential energy. When the mass is allowed to fall the potential energy is converted to kinetic energy. The further the mass falls, the smaller the value of potential energy and the greater the value of kinetic energy. As the mass falls, the velocity increases and so does the kinetic energy, although the increase varies as the square of the velocity (see equation (5.10)).

It can, therefore, be shown that for a given total energy value and in the absence of friction:

$$\text{loss of potential energy} = \text{gain in kinetic energy} \qquad (5.6)$$

and the opposite is also true:

$$\text{gain in potential energy} = \text{loss in kinetic energy} \qquad (5.7)$$

### Example 5.7

A mass of 10 kg is raised through a height of 30 m. Find the potential energy imparted to the mass.

*Solution*

Since work has to be performed to impart potential energy to the mass, equation (5.1) can be used:

$$w = F \times s$$

Now $s = 30\,\text{m}$ and $F = mg = 10 \times 9.81 = 98.1\,\text{N}$, so that

$$w = 98.1 \times 30 = 2943\,\text{J}$$

and is the work done in raising the mass to a height of 30 m. This is the potential energy possessed by the mass at that height.

### 5.3.2 Kinetic energy

Kinetic energy is energy possessed by a body by virtue of its velocity. Consider how that velocity may be achieved by studying Newton's second law and the equation (3.2)

$$F = ma$$

which can be considered to be the constant magnitude force $F$ required to accelerate a mass $m$, from a lower velocity to a higher velocity. In increasing the speed, work must have been done by the force by moving the mass through a distance. By multiplying equation (3.2) by the distance $s$, an equation for work can be derived as follows:

$$Fs = mas$$

or

$$w = mas$$

or

$$KE = mas \tag{5.8}$$

Now equation (2.5) can be transposed to isolate $s$ as follows:

$$v_2^2 = v_1^2 + 2as$$

giving

$$s = \frac{v_2^2}{2a} \tag{5.9}$$

where the initial velocity $v_1$ is assumed to be zero.
  Inserting equation (5.9) into equation (5.8) gives

$$KE = \frac{mav_2^2}{2a}$$

so that

$$KE = \tfrac{1}{2}mv^2 \quad \text{(J)} \tag{5.10}$$

  It may be noted that the work done on a body is equal to the kinetic energy gained by that body.

### Example 5.8

Find the kinetic energy of a mass of 2000 kg, moving with a velocity of 40 km/h.

*Solution*

The kinetic energy can be found by using equation (5.10):

$$KE = \tfrac{1}{2}mv^2$$

Now $m = 2000\,\mathrm{kg}$ and

$$v = \frac{40 \times 1000}{60 \times 60} = 11.1\,\mathrm{m/s}$$

so that

$$KE = \frac{2000 \times 11.1^2}{2} = 123\,000\,\mathrm{J}$$

$$= 123\,\mathrm{kJ}$$

### 5.3.3 Kinetic energy applied to angular motion

The concept of kinetic energy can also be applied to rotating objects. The same proof applies as given in section 5.3.2, providing that equivalent units are used. Thus equation (5.10) becomes

$$KE = \tfrac{1}{2}I\omega^2 \quad \text{(J)} \tag{5.11}$$

where $I$ = moment of inertia $(\mathrm{kg/m^2})$,
$\omega$ = angular velocity $(\mathrm{rad/s})$.

One of the most useful applications of rotational kinetic energy is that of flywheels used in reciprocating engines, such as steam or internal combustion engines. At low speeds the flywheel stores enough kinetic energy to carry the mechanism back to the beginning of the cycle, ready for another injection of energy. At high speeds the flywheel receives and donates energy and in so doing, smooths out the energy pulses applied to the crank from the pistons. Another modern application of angular kinetic energy is where a large flywheel is installed under the floor of a passenger bus. When the bus slows, the linear kinetic energy possessed by the bus is absorbed by the flywheel which increases in speed. When it is necessary for the bus to accelerate, the energy stored in the flywheel is mechanically diverted to the driving wheels. The whole process stores and reuses energy which would normally be lost in braking.

### Example 5.9

Find the kinetic energy of a rotor of mass $200\,\mathrm{kg}$ and radius of gyration of $250\,\mathrm{mm}$, when rotating at $600\,\mathrm{rev/min}$.

*Solution*

The kinetic energy can be found using equation (5.11):

$$KE = \tfrac{1}{2}I\omega^2$$

Now $I = mk^2 = 200 \times 0.25^2 = 12.5\,\mathrm{kg\,m^2}$ and

$$\omega = \frac{600 \times 2\pi}{60} = 62.8\,\mathrm{rad/s}$$

so that

$$KE = \frac{12.5 \times 62.8^2}{2} = 24\,600\,J$$

$$= 24.6\,kJ$$

### 5.3.4  Total kinetic energy of a rotating body

When a wheel rolls along the ground it possesses kinetic energy due to its angular velocity and also kinetic energy due to its linear velocity. It can be said, therefore, that the total kinetic energy of any rotating body is made up of two parts:

1. that based upon linear motion as described in equation (5.10):

   kinetic energy $= \frac{1}{2}mv^2$

2. that based upon angular motion as described in equation (5.11):

   kinetic energy $= \frac{1}{2}I\omega^2$

For most practical cases the centre of mass coincides with the centre of rotation and when this occurs the total kinetic energy can be described as follows:

total kinetic energy $= \frac{1}{2}mv^2 + \frac{1}{2}I\omega^2$         (5.12)

### Example 5.10

A flywheel consists of a uniform steel disc of mass 20 kg and diameter 0.4 m. It is lifted vertically 2.0 m, and placed on an incline. The flywheel is then released and rolls down the incline without slipping. Find:

(a) the moment of inertia of the flywheel;
(b) the potential energy;
(c) the maximum linear velocity of the disc as it reaches the bottom of the incline;
(d) the maximum angular velocity.

### Solution

(a) The moment of inertia can be found using equation (3.8):

$$I = mk^2$$

Now $r = 0.2$ m, $m = 20$ kg and

$$k = \frac{r}{\sqrt{2}} = \frac{0.2}{\sqrt{2}} = 0.14\,m$$

so that

$$I = 20 \times 0.14^2 = 0.4\,kg\,m^2$$

(b) Since work has to be performed to impart potential energy to the disc, equation (5.1) can be used:

potential energy $= w = F \times s$

Now $s = 2.0\,\text{m}$ and $F = 20 \times 9.81 = 196.2\,\text{N}$ so that

$w = 196.2 \times 2 = 392.4\,\text{J}$

(c) The maximum velocity can be found using equation (5.12):

total kinetic energy $= \frac{1}{2}mv^2 + \frac{1}{2}I\omega^2$

This equation can be manipulated to eliminate $\omega$, by using equation (2.15):

$v = \omega r$

or

$$\omega = \frac{v}{r} \qquad \text{(i)}$$

Inserting equation (i) into equation (5.12) gives

total kinetic energy $= \frac{1}{2}mv^2 + \frac{1}{2}\frac{Iv^2}{r^2}$

$$= \frac{1}{2}\left(mv^2 + \frac{Iv^2}{r^2}\right) \qquad \text{(ii)}$$

As the disc reaches the bottom of the slope all the potential energy has been converted to kinetic energy and hence equation (5.6) can be used:

loss of potential energy $=$ gain in kinetic energy

Inserting equation (ii) above

potential energy $= \frac{1}{2}\left(mv^2 + \frac{Iv^2}{r^2}\right)$

Now potential energy $= 392.4\,\text{J}$, $m = 20\,\text{kg}$, $I = 0.4\,\text{kg}\,\text{m}^2$ and $r = 0.2\,\text{m}$, so that

$$392.4 = \frac{1}{2}\left(20v^2 + \frac{0.4v^2}{0.2^2}\right)$$

$$= \frac{20v^2 + 20v^2}{2}$$

$$v = \sqrt{\frac{392.4}{20}} = 4.43\,\text{m/s}$$

(d) The maximum angular velocity can be found by using equation (2.15):

$v = \omega r$ or $\omega = \frac{v}{r}$

Now $v = 4.43\,\text{m/s}$ and $r = 0.2\,\text{m}$, so that

$$\omega = \frac{4.43}{0.2} = 22.15\,\text{rad/s}$$

### 5.3.5 The principle of the conservation of energy

Energy cannot be created or destroyed. It can exist in a variety of forms such as mechanical, electrical, chemical or heat energy, but a loss of energy in any one form is always accompanied by an equivalent gain in energy in another form.

Consider the chemical energy stored in petrol. When this is burned, in an internal combustion engine, a considerable amount of the energy is converted into heat and is subsequently passed to the surroundings through the engine cooling system. The bulk of the remaining energy is converted into mechanical energy and is used to drive the car and its various systems.

A body which slides along a rough surface is slowed by the sliding action, losing kinetic energy as it does so. The kinetic energy is lost due to the work done against friction forces. The 'lost' energy may be lost for mechanical reasons but it has been converted to heat energy. An excellent example of this is the vehicle disc brake whose temperature increases as it absorbs energy in slowing the vehicle.

More generally, when a body falls it loses potential energy, but gains an equivalent amount of kinetic energy owing to its increase in velocity. When the body reaches the ground all the potential energy has been converted to kinetic energy.

## 5.4 POWER

When describing the energy output of, say, an internal combustion engine or a turbine, the word 'power' is often loosely applied. It may be said that the engine is 'powerful', but this refers to mechanical output power.

The term 'power' may be considered in a number of different ways and depends upon how the power is generated, stored and used. For instance, chemical 'power' is stored as petrol and is released when petrol is burned. The rate of heat generated may be measured in watts (the unit of power). Petrol, of course, is burned in an internal combustion engine, which has a mechanical output power, also measured in watts.

Electricity is generated by forcing high-temperature steam through a turbine. The 'power' possessed by the steam turns the turbine which, in turn, drives the generator, thus producing the electricity. This can also be measured in watts. The electrical power can be used to turn an electric motor which converts it back into mechanical power.

Whatever the means of conversion or generation, the power developed is measured in watts, and if losses are ignored, the power possessed by one medium can be converted to, and used in, another medium.

Power is the rate of doing work. The base unit of power is the joule/s, designated J/s, and, in mechanical terms, can be defined by considering: 1 N moved through 1 m in 1 s, so that the unit of power is J/s = W.

This can be explored further by considering equation (5.1):

$$w = F \times s$$

Dividing both sides by time gives

$$\frac{w}{t} = F \times \frac{s}{t} \qquad (5.13)$$

but

$$\frac{s}{t} = \text{velocity, } v \quad \text{and} \quad \frac{w}{t} = \text{power, } P$$

so that equation (5.13) can be rewritten:

$$w = F \times v \quad \text{J/s} \quad \text{or} \quad \text{W} \qquad (5.14)$$

Convention dictates that the unit of J/s is redesignated as the watt or W.

The watt is often found to be too small to be used effectively and so other derivatives are used such as:

kilowatt = 1000 W
megawatt = 1 000 000 W

Often the amount of power used is measured for a period of time. For instance, domestic electricity is measured in kilowatt-hours.

$$\begin{aligned} 1 \,\text{kWh} &= 1000 \,\text{Wh} \\ &= 1000 \times 3600 \,\text{Ws} \\ &= 3\,600\,000 \,\text{J} \\ &= 3.6 \,\text{MJ} \end{aligned}$$

and is the work done or energy expended in one hour.

### 5.4.1 Power related to rotation

It is often the case that generated mechanical power is transmitted by means of a rotating shaft. The power output from this can be quantified in terms of watts. Equation (5.5) shows

$$w = 2\pi T n \quad \text{J}$$

Since power is the rate of doing work, equation (5.5) can be divided by time to a base of 1 s, thus

$$w = \frac{w}{t} = \frac{2\pi T n}{60} \qquad (5.15)$$

where $T$ = torque (N m),
$n$ = rev/min.

Equation (5.15) should be considered further.

$$\text{angular velocity} = \omega = \frac{2\pi n}{60} \quad \text{rad/s}$$

thus equation (5.14) can be rewritten

$$w = T\omega \tag{5.16}$$

### Example 5.11

A car pulls a trailer at 80 km/h when exerting a steady pull of 900 N. Calculate:

(a) the work done in 20 min;
(b) the power required at any instant.

*Solution*

(a) The work done in 20 min can be found by using equation (5.1):

$$w = F \times s$$

Now $F = 900\,\text{N}$ and

$$s = 80 \times 1000 \times \frac{20}{60} = 26\,667\,\text{m}$$

so that

$$w = 900 \times 26\,667 = 24\,\text{MJ}$$

(b) The power required at any instant can be found by using equation (5.15):

$$w = \frac{w}{t}$$

Now $w = 24\,\text{MJ}$ and $t = 20\,\text{min} = 20 \times 60 = 1200\,\text{s}$, so that

$$w = \frac{24\,000\,000}{1200} = 20\,\text{kW}$$

### Example 5.12

A flywheel, of mass 500 kg, takes the form of a uniform disc of diameter 1.5 m. The flywheel is fastened to a horizontal shaft lying in bearings and is coupled to an electric motor. Find:

(a) the moment of inertia of the flywheel assuming the contribution of the shaft is negligible;
(b) the torque at the motor to give the flywheel/shaft system an acceleration of 6 rad/s$^2$;
(c) the power transmitted by the motor when a speed of 500 rev/min has been attained;

(d) the height a mass of 2000 kg will be raised before the flywheel comes to rest. The mass is fastened to a rope which is wrapped around a drum, which, in turn, is keyed to the flywheel shaft. When the flywheel reaches a speed of 650 rev/min the motor is disconnected and the flywheel/drum system is allowed to rotate freely.

*Solution*

(a) The moment of inertia can be found by using equation (3.7):

$$I = mk^2 \quad \text{where } k = \frac{r^2}{2}$$

Now $m = 500$ kg and

$$r = \frac{1.5}{2} = 0.75 \text{ m}$$

so that

$$I = \frac{500 \times 0.75^2}{2} = 140.63 \text{ kg m}$$

(b) The torque required to give the system an acceleration of 6 rad/s² can be found using equation (3.5):

$$T = I\alpha$$

Now $I = 140.63$ kg m² and $\alpha = 6$ rad/s², so that

$$T = 140.63 \times 6 = 843.78 \text{ N m}$$

(c) The power transmitted by the motor can be found by using equation (5.15):

$$w = \frac{2\pi Tn}{60}$$

Now $T = 843.78$ N m and $n = 500$ rev/min, so that

$$w = \frac{2\pi \times 843.78 \times 500}{60} = 44\,180 \text{ W}$$
$$= 44.18 \text{ kW}$$

(d) The height a mass of 500 kg can be raised can be found by first calculating the kinetic energy stored in the flywheel, by using equation (5.11):

$$KE = \tfrac{1}{2}I\omega^2$$

Now $I = 140.63$ kg m² and

$$\omega = \frac{650 \times 2\pi}{60} = 68 \text{ rad/s}$$

so that

$$KE = \frac{140.63 \times 68^2}{2} = 325.1 \text{ kJ}$$

The height a mass can be raised can now be found by considering equation (5.7):

gain in potential energy = loss of kinetic energy

or

$mgh = KE$

where $h$ = height (m). Now $m = 2000\,kg$, $g = 9.81\,m/s^2$ and $KE = 325.1\,kJ$, so that

$$2000 \times 9.81 \times h = 325.1 \times 10^3$$

$$h = \frac{325.1 \times 10^3}{2000 \times 9.81} = 16.56\,m$$

### 5.4.2 Efficiency of a machine

Whenever energy is converted from one form to another, it is a fact of life that some losses occur. For example, the chemical energy stored in petrol is released when petrol is burned in the internal combustion engine; however, some of the energy is unavoidably lost as heat through the walls of the engine. Furthermore, energy is required to move and turn the necessary parts of the engine and gearbox, so that the energy stored in the petrol is somewhat diminished by the time it is made available at the wheels.

When the energy (or power) output is compared with the energy (or power) input, an efficiency rating can be calculated, as follows:

$$\text{efficiency} = \frac{\text{output power}}{\text{input power}} \qquad (5.17)$$

Efficiency is normally expressed as a percentage, and is calculated as:

$$\text{efficiency} = \eta = \frac{\text{output power} \times 100}{\text{input power}} \ \% \qquad (5.18)$$

### Example 5.13

An electric motor/gearbox drive system has a rated electrical input power of 75 kW and a measured torque output of 1910 N m at a speed of 300 rev/min. Find:

(a) output power;
(b) the percentage efficiency of the system.

*Solution*

(a) The output power can be found by using equation (5.15):

$$w = \frac{2\pi Tn}{60}$$

Now $T = 1910\,\mathrm{N\,m}$, $n = 300\,\mathrm{rev/min}$, so that

$$w = \frac{2\pi \times 1910 \times 300}{60} = 60\,\mathrm{kW}$$

(b) The efficiency as a percentage of overall power can be found using equation (5.18):

$$\text{efficiency} = \eta = \frac{\text{output power} \times 100}{\text{input power}}$$

Now output power $= 60\,\mathrm{kW}$, input power $= 75\,\mathrm{kW}$, so that

$$\eta = \frac{60 \times 100}{75} = 80\%$$

## 5.5 SUMMARY

Key equations that have been introduced in this chapter are as follows.
Work done:

$$w = F \times s \quad \text{(J)} \tag{5.1}$$

or

$$w = F_{AV} \times s \quad \text{(J)} \tag{5.2}$$

$$w = T \times \theta \quad \text{(J)} \tag{5.4}$$

$$w = 2\pi Tn \quad \text{(J)} \quad \text{for } n \text{ rev} \tag{5.5}$$

$$\text{loss of potential energy} = \text{gain in kinetic energy} \tag{5.6}$$

$$\text{gain in potential energy} = \text{loss in kinetic energy} \tag{5.7}$$

Kinetic energy for linear motion:

$$KE = \tfrac{1}{2}mv^2 \quad \text{(J)} \tag{5.10}$$

Kinetic energy for angular motion:

$$KE = \tfrac{1}{2}I\omega^2 \quad \text{(J)} \tag{5.11}$$

Total kinetic energy:

$$KE = \tfrac{1}{2}mv^2 + \tfrac{1}{2}I\omega^2 \quad \text{(J)} \tag{5.12}$$

Power for linear motion:

$$P = F \times v \quad \text{(W)} \tag{5.14}$$

Power for angular motion:

$$P = \frac{w}{t} = \frac{2\pi Tn}{60} \tag{5.15}$$

$$P = T\omega \tag{5.16}$$

Efficiency:

$$\eta = \frac{\text{output power} \times 100}{\text{input power}} \quad \% \tag{5.18}$$

## 5.6 PROBLEMS

1. A force of $F$ newtons acting on a body in the direction of its motion varies as follows for different distances $s$ from the start point:

   $$\begin{array}{rcccccc}
   F = & 200 & 380 & 470 & 500 & 420 & 320 \\
   s = & 0 & 10 & 20 & 30 & 40 & 50
   \end{array}$$

   Draw the work diagram to scale and find:
   (a) the average force;
   (b) the work done in kJ when the body is moved through a distance of 60 m.
2. The output power of an electric motor is 10 kW and is maintained constant for 7 h. Find:
   (a) the work done in kWh and in MJ;
   (b) the value of the torque if the speed of the motor is 600 rev/min.
3. A belt friction brake applied at the circumference of a pulley of 450 mm diameter exerts a backward drag of 250 N. If the speed of the pulley is 680 rev/min, find:
   (a) the torque applied to the pulley;
   (b) the power absorbed by the brake;
   (c) the efficiency of the motor if it has an input power of 4.3 kW.
4. The diesel engine of a 500 t train increases the speed of the train uniformly from rest to 15 m/s in 200 s, along a horizontal track. Find the average power developed.
5. An motor car, having a mass of 2000 kg, travels up an 8° slope at a constant speed of $v = 90$ km/h. If mechanical resistance and wind resistance are neglected, find the power developed by the engine if the overall efficiency is $\eta = 0.7$.
6. A spring, having a stiffness of 6 kN/m, is compressed 500 mm. The stored energy in the spring is used to drive a machine which requires 80 W. Determine how long the spring can supply energy at the required rate.
7. An electric motor is used to hoist a loaded industrial elevator upward with a constant velocity of 9 m/s. If the motor draws 65 kW of electrical power and the total mass of the elevator is 550 kg, find the efficiency of the motor, neglecting the effect of the mass of any of the mechanism parts.

# Dry friction | 6

## 6.1 AIMS

- To introduce dry friction and explain its causes.
- To explain how dry friction affects mechanical systems.
- To analyse the friction forces associated with bodies moving on flat, horizontal surfaces.
- To analyse the friction forces associated with bodies moving on inclined planes.
- To analyse the effect of dry friction when applied to screw threads.

## 6.2 INTRODUCTION TO FRICTION

The term 'friction' describes the effect of resistance to motion when two contacting surfaces attempt to move relative to each other. At a microscopic level the contacting surfaces touch at a relatively small number of points. As the surfaces slide over each other, resistance to motion results from the interference and 'catching' of the contact points. Figure 6.1 shows an enlarged microscopic contact region.

The surface texture affects the value of the frictional resistance and no matter how smooth the surfaces there will always be frictional resistance present. This can be used to great advantage when considering such things as walking, hanging on a rope, a screw jack, vehicle traction on the

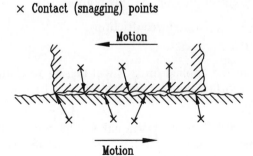

**Fig. 6.1** Enlarged microscopic contact region.

road or a vehicle clutch. In all of these cases lack of friction would allow the contact surfaces to slip. The presence of friction resists the slipping effect and permits the actions to take place.

There are some situations where friction causes energy to be consumed and components to wear. When two surfaces rub together, energy is used in overcoming the frictional resistance. This often results in heat being generated. This can be observed when rubbing hands together to keep warm. Due to the high loads and speeds within some machinery, the heat generated can be very high and can cause damage to the components. The energy used in such an action is lost to the surroundings as generated heat, thus reducing the general efficiency of the machine.

Two contacting, sliding surfaces will also wear. The microscopic contact points will often break off, especially with rough surfaces, effectively abrading each surface and generating sharp debris to cause damage elsewhere in the system. When heat and wear take place together the surfaces 'pick up' melted debris from the adjacent faces, causing very rough surfaces which, in turn, cause even more damage.

In order to overcome a large proportion of these effects it is possible to introduce lubricants between the mating faces, thus allowing sliding to take place. By doing this, there is a substantial reduction in wear and energy consumption. The introduction of lubricants separates the two surfaces, allowing them to 'ride' on a cushion of oil or grease and largely prevents contact.

In general two types of friction can occur between surfaces. **Fluid friction** exists when the two surfaces are separated by a fluid film and is dependent upon the fluid's ability to shear. A thick fluid, such as treacle, is said to have a high viscosity and is difficult to shear. A fluid of low viscosity, such as water, is easy to shear. Since fluid friction is studied later, in Chapter 14, only the effects of dry friction are discussed here.

**Dry friction**, sometimes called Coulomb friction after the engineer and physicist C. A. Coulomb, occurs between contacting surfaces in the absence of any lubricating fluid.

An appreciation of the effects of friction may be gained by considering a block resting on a horizontal surface, Figure 6.2. The block applies a normal load $N$ to the surface due to gravity acting on its mass. The force $P$ is necessary to move the block along the surface and increases until the block is on the point of moving. At this point

applied force = resisting force (friction)
$$P = F_r$$

Any addition to the applied force $P$ will merely accelerate the block. There are four important concepts which relate to friction and these are considered below.

**Fig. 6.2** Block resting on a horizontal surface.

1. The friction force $F_r$ is proportional to the normal load $N$. This concept is true only within the limits indicated in Figure 6.3, which shows the relationship between $F_r$ and $N$. The useful portion is limited to the straight line section of the curve. Since $F_r$ and the normal force

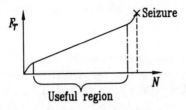

**Fig. 6.3** The limits of the linear relationship between the friction force and the normal load.

$N$ are proportional, their relationship is constant and can be written:

$$\frac{F_r}{N} = \mu \tag{6.1}$$

where $\mu$ is the coefficient of friction. This can be arranged to give the basic equation:

$$F_r = \mu N \tag{6.2}$$

When the block is on the point of moving the applied force is

$$P = F_r = \mu N \tag{6.3}$$

2. The friction force is independent of area of contact between the surfaces. In order to understand how the friction force can be independent of area, consideration must be given to the source of friction. It is generally understood that dry friction results from point contact between the microscopically rough surfaces, as shown in Figure 6.1. When on the point of moving, the applied force $P$ is the same as the friction force $F_r$ and it is this force which is needed to move the block over the contact points.

   Increasing the area creates more contact points and proportionally distributes the normal force $N$. Hence if $N$ remains unchanged, then,

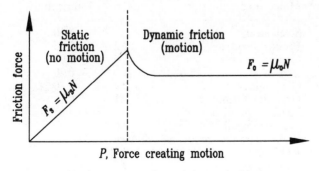

**Fig. 6.4** The relationship between static and dynamic friction.

no matter what the area, the friction force will also remain unchanged. Since the friction force remains unchanged it follows that the coefficient of friction, $\mu$, also remains unchanged.

3. The friction force is independent of the speed of slipping, within limits. The force required to initiate movement is always greater than the force required to keep the object sliding. This observation introduces the concepts of static and dynamic friction. Figure 6.4 shows the relationship between $F_r$ and the applied force. $P$ before and after motion commences. The dynamic friction force is often considered to be approximately 75% of the static friction force value and, within limits, is independent of the speed of sliding. The limits referred to are: very slow moving, such as the transition between stationary and just moving, and very high speeds where excessive energy dissipation results in the surfaces overheating, possibly leading to fusion. The technique of friction welding uses this principle.

4. The friction force is dependent upon the nature of the surfaces in contact. Materials and their relative roughness will affect the value of the frictional resistance. Rough surfaces will make contact as described in Figure 6.1. However, the surfaces will tend to interlock with increasing surface roughness and hence increase the value of frictional resistance. The coefficient of friction will also be proportionally increased. Smooth surfaces, however, do not eradicate the effects of friction for, no matter how smooth the surfaces, there will always be frictional resistance present.

The frictional resistance and hence the coefficient of friction will also vary between materials whether they are like materials or different materials in contact. Typical values of the coefficient of friction for different materials in contact are presented in Table 6.1.

**Table 6.1** Table of average values of coefficients of friction for different materials

| Material combination | Coefficient of friction | |
|---|---|---|
| | Dry | Lubricated |
| Steel–steel | 0.7 | 0.10 |
| Bronze–bronze | 0.2 | 0.05 |
| Copper–copper | 1.0 | 0.08 |
| Iron–iron | 1.0 | 0.20 |
| Glass–glass | 0.9 | 0.30 |
| Graphite–Graphite | 0.1 | 0.10 |
| Leather–wood | 0.3 | – |
| Leather–metal | 0.6 | – |
| Brake material–cast iron | 0.4 | – |
| PTFE–steel | 0.05 | – |
| Nylon–steel | 0.4 | – |

## 6.3 LUBRICATION AND FLUID FRICTION

Although fluid friction is a topic that is discussed in Chapter 14 it is briefly mentioned here to illustrate the influence of a lubricant on friction. When a fluid is introduced between two surfaces as a lubricant, it can be assumed that the surfaces are no longer in contact and any friction is due to the shearing of the fluid.

The consistency of a fluid, whether thick or thin, is known as its viscosity. The larger the value of viscosity, the harder it is to shear the fluid. This leads to a larger value of friction and it becomes increasingly more difficult for the fluid to move. To appreciate this concept, consider the behaviour of treacle and that of water.

Lubricants may be in the form of oils, semi-solids such as grease or true solids such as graphite or Teflon. Many other substances may be used and are usually matched to an application by considering factors such as load, speed, temperature, viscosity and environment.

## 6.4 STICTION

This is a hybrid term describing the action of two adjacent and moving surfaces appearing to stick together, releasing, sticking and releasing in quick succession. This often happens so fast that it can be a source of vibration. For instance, push a damp finger across a desk and note the rubbing sound.

In industrial applications this phenomenon is mostly confined to elastic components such as seals, where the elasticity in the seal allows the surfaces to 'stick' together. As the applied load $P$ stretches the rubber seal, the frictional resistance increases until it is finally overcome. The rubber, being in a stretched state, then tries to regain its unstretched condition, thus slipping over the surface. Figure 6.5 shows the sequence. Stiction often occurs in the transition between stationary and dynamic states and

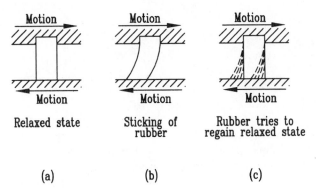

Fig. 6.5 The mechanism by which stiction occurs.

can happen at both relatively high frequencies and slow speeds. It is dependent upon a number of physical factors and is difficult both to define and to quantify. Stiction is often found to be a problem in slow-moving hydraulic rams.

## 6.5 FRICTION ON A HORIZONTAL PLANE

Consideration of a block resting on a horizontal surface is usually a suitable approximation for considering many problems. This allows the use of equation (6.3), i.e.

$$P = F_r = \mu N$$

It must be stressed that the applied force $P$ equals the friction force $F_r$ only when the block is on the point of moving. Any increase in $P$ after motion commences will merely accelerate the block.

**Example 6.1**

A wooden packing-case weighing 2000 N lies on a horizontal wooden floor. It is pushed across the floor by a man applying a horizontal force of 800 N. Determine the coefficient of friction.

$$N = 2000\,\text{N}$$
$$P = 800\,\text{N}$$
$$\mu = ?$$

*Solution*

The coefficient of friction is the result of the ratio $P/N$ when on the point of moving and in this state $P = F_r$. Equation (6.3) may be rearranged:

$$P = F_r = \mu N$$

$$\frac{P}{N} = \mu$$

$$\mu = \frac{800}{2000} = 0.4$$

**Example 6.2**

The wooden packing-case in the previous example has been filled with components and has a mass of 356.8 kg. The packing-case is to be pushed to a loading bay across a concrete floor with a coefficient of friction of 0.3. If two men now push the case with a horizontal force of 1600 N, determine the excess force available after friction has been overcome.

*Solution*

In this case the applied force is not the same as the friction force. After friction has been overcome the excess force is available to move the packing-case.

$$F_r = ?$$
$$P = 1600\,\text{N}$$
$$\mu = 0.3$$
$$N = 356.8 \times 9.81 = 3500\,\text{N}$$

The frictional resisting force can be found using equation (6.3):

$$F_r = \mu N$$
$$= 0.3 \times 3500 = 1050\,\text{N}$$

The excess force is the difference between the applied force and the friction force:

$$\text{excess force} = P - F_r$$
$$= 1600 - 1050 = 550\,\text{N}$$

### 6.5.1 Friction on a horizontal plane with an inclined applied force

In practice $P$ is often inclined to the horizontal friction plane as shown in Figure 6.6. In this case the force which moves the block along the plane is the horizontal component of $P$, $P_H = P\cos\theta$.

The vertical component of $P$, i.e. $P_v = P\sin\theta$ tends to lift the block and hence reduces the normal load $N$.

### Example 6.3

A box sled full of castings is pulled along a horizontal concrete floor by means of a rope attached to the front of the sled. The rope is inclined at $30°$ to the horizontal (Figure 6.7).

(a) Determine the horizontal and vertical components of the applied force if the tension in the rope is 1000 N.

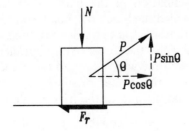

**Fig. 6.6** Idealized block on a horizontal friction plane with an inclined applied load.

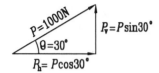

**Fig. 6.7** Horizontal and vertical components of the applied inclined force.

(b) Find the resistance to motion due to friction if the coefficient of friction is 0.2 and the mass of the box complete with components is 300 kg.

*Solution*

(a) The horizontal and vertical components can be found by applying trigonometry to Figure 6.7:

$$P_H = P \cos \theta$$
$$= 1000 \cos 30° = 866 \, N$$
$$P_v = P \sin \theta$$
$$= 1000 \sin 30° = 500 \, N$$

(b) To find the frictional resistance it is first necessary to find $N$; $N$ comprises:

force acting down due to the weight $= 300 \times 9.81$
$$= 2943 \, N$$

force acting upwards due to $P = P \sin \theta = 500 \, N$

so that

$$N = 2943 - 500 = 2443 \, N$$

$F_r$ can now be found using equation (6.3):

$$F_r = \mu N$$
$$= 0.2 \times 2443 = 488.6 \, N$$

## 6.6 FRICTION ON AN INCLINED PLANE

If a block is placed on an inclined plane, it is influenced by the friction force $F_r$ and its own weight $W$ as shown in Figure 6.8, in which $W =$ weight and $\theta° = \delta°$ for impending motion.

### 6.6.1 Angle of friction

When the plane is inclined at a shallow angle $\theta$ the block will remain stationary. As the inclination of the plane is increased, it will reach a particular angle at which the block will just begin to slide. This angle is termed the 'angle of friction' although it is sometimes called the 'angle of repose', and it occurs when the forces acting parallel to, and down the plane equal the friction force which acts up the plane.

Figure 6.8 shows how the weight of the block can be resolved into components parallel and normal to the friction plane:

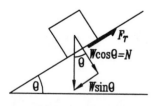

**Fig. 6.8** A block placed on an inclined plane.

- $W \sin \theta$ is parallel to the plane and is the component which always tries to pull the block down the plane.
- $W \cos \theta$ is at 90° to the plane and can be considered as the normal force $N$.

When $W \sin \theta$ equals $F_r$ the condition is that of impending motion and the block is on the point of sliding. At this point:

the angle of inclination $\theta° = \delta°$, the angle of friction

It is also true that

$$W \sin \delta = F$$
$$= \mu N$$
$$= \mu W \cos \delta \qquad (6.4)$$

and can be written

$$\frac{W \sin \delta}{W \cos \delta} = \mu$$

or expressed as

$$\tan \delta = \mu \qquad (6.5)$$

### Example 6.4

A company wishes to install a steel ramp which will allow cardboard boxes to slide from an upper to a lower level. Calculate the angle at which the ramp should be fixed so that the boxes will be on the point of sliding. Each box has a mass of 5 kg and the coefficient of friction between cardboard and steel is 0.4 (Figure 6.9).

*Solution*

The angle can be found using equation (6.5):

$$\mu = \tan \delta$$
$$\delta = \tan^{-1}(\mu)$$
$$\delta = \tan^{-1}(0.4) = 21.8°$$

It is important to emphasize that the angle of friction is a unique angle of inclination for a particular pair of materials in contact since it is this angle which gives the condition of impending motion.

### 6.6.2  Forces applied to a body lying on an inclined plane

If the angle of inclination of a plane is always greater than the angle of friction, then the block, if left, will tend to slide down the plane under its own weight.

The friction force can be calculated by using equation (6.3) and referring to Figure 6.8.

$$F_r = \mu N$$
$$= \mu W \cos \theta \qquad (6.6)$$

where $N = W \cos \theta$.

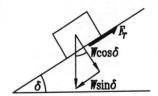

**Fig. 6.9** Idealization of a cardboard box on an inclined plane.

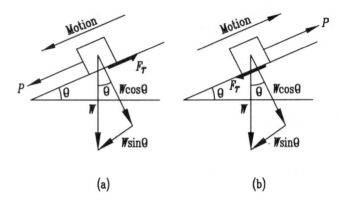

**Fig. 6.10** A body lying on an inclined plane.

A body lying on an inclined plane may move either upwards or downwards depending on the direction of the applied force. Figure 6.10(a) shows that the body will slide down the plane when the force is applied in that direction. Note that the friction force opposes motion by acting up the plane. Conversely, Figure 6.10(b) shows that when the force is applied up the plane the body will move in that direction. The friction force again opposes motion by acting down the plane.

As a general case, forces on the inclined plane may be resolved into those acting up the plane and those acting down the plane. If a block placed on an inclined plane is on the point of moving, either up or down, then the forces acting on the block must be holding it in that condition. Put another way, all the forces acting up the plane must equal all the forces acting down the plane to ensure equilibrium.

No matter how the forces may be applied to a block, this basic assumption is true whenever the block is on the point of moving. The general approach is, therefore, to resolve all forces parallel to the friction plane and equate all those acting up the plane to all those acting down.

There follow a number of examples which outline the basic principles involved in handling the various forces which may be applied to a body lying on an inclined plane.

### 6.6.3 Forces on a body moving up an inclined plane

The force required to overcome friction and resistance due to gravity is $P$ and is applied up the plane as shown in Figure 6.10(b). When on the point of moving, the forces acting up the plane must equal the forces acting down the plane and can be equated as follows:

forces acting up the plane = forces acting down the plane

$$P = W \sin \theta + F_r$$
$$= W \sin \theta + \mu W \cos \theta \tag{6.7}$$

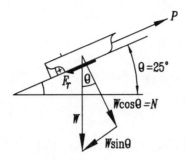

**Fig. 6.11** Boat being winched up a slipway.

### Example 6.5

A 1500 kg boat is winched up a slip inclined at 25° (Figure 6.11). If $\mu = 0.5$, find the force in the winch cable which is parallel to the slip.

*Solution*

The force in the winch cable can be found by using equation (6.7). The condition is motion up the plane.

$$W = 1500 \times 9.81 = 14\,715\,\text{N}$$
$$P = W \sin\theta + \mu W \cos\theta$$
$$= 14\,715 \times \sin 25° + 0.5 \times 14\,715 \times \cos 25°$$
$$= 6218.8 + 6668.2$$
$$= 12\,887\,\text{N} \quad \text{or} \quad P = 12.9\,\text{kN}$$

#### 6.6.4 Forces on a body moving down an inclined plane

In this case the applied force and the force due to gravity are acting down the plane, while friction, which opposes motion, acts up the plane as shown in Figure 6.10(a). When on the point of moving the forces acting up the plane must equal the forces acting down the plane and can be equated as follows:

forces acting up the plane = forces acting down the plane

$$F_r = W \sin\theta + P \tag{6.8}$$
$$\mu W \cos\theta = W \sin\theta + P \tag{6.9}$$

### Example 6.6

A casting of mass 2000 kg is to be hauled down a slope inclined at 15°, by a steel cable running parallel to the slope. If $\mu = 0.4$, find the force $P$ required to pull the casting down the slope.

$$\theta = 15$$
$$\mu = 0.4$$
$$W = 2000 \times 9.81 = 19\,620\,\text{N}$$

*Solution*

The force required to pull the casting down the slope can be found by using equation (6.9):

$$\mu W \cos \theta = W \sin \theta + P$$

Rearranging

$$P = \mu W \cos \theta - W \sin \theta$$
$$= 0.4 \times 19\,620 \times \cos 15° - 19\,620 \times \sin 15°$$
$$= 2502\,\text{N}$$

### 6.6.5 Oblique forces applied to a body on an inclined plane

When a force is applied to a body which lies on an inclined plane, the force may not lie parallel to the plane, as can be seen in Figure 6.12. The force may either be moving the body up the plane or it may be moving it down the plane. In both cases the force needs to be resolved into components which lie normal to the plane and parallel to the plane. In Figure 6.12 $\theta$ = angle of the inclined plane and $\phi$ = angle of the applied force to the plane.

The procedure for analysis has previously been explained. The forces acting up the plane are equated to all the forces acting down the plane, as follows:

forces acting up the plane = forces acting down the plane

In this case, however, the analysis is slightly more complex since the components which lie at 90° to the plane have to be included. The method differs slightly depending upon whether the motion is up the plane or down the plane.

It can be seen, in Figure 6.12(a), that motion up the plane requires that the components of the oblique force are applied as follows:

up and parallel to the plane = $P_p = P \cos \phi$
into and normal to the plane = $P_n = P \sin \phi$

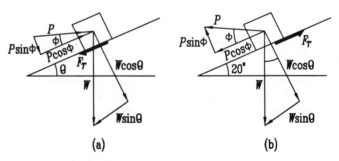

Fig. 6.12 Oblique forces applied to a body lying on an inclined plane.

Of these, $P\cos\phi$ is the only component which can push the block up the plane, and $P\sin\phi$ actually adds to the normal force component $W\cos\theta$. The forces can be equated as follows:

forces acting up the plane = forces acting down the plane

$$P\cos\phi = F_r + W\sin\theta$$
$$= \mu N + W\sin\theta$$
$$= \mu(W\cos\theta + P\sin\phi) + W\sin\theta \qquad (6.10)$$

It may be observed in Figure 6.12(b) that motion down the plane requires that the components of the oblique force are applied as follows:

down and parallel to the plane $= P_p = P\cos\phi$

out of and normal to the plane $= P_n = P\sin\phi$

It can be seen that $P\cos\phi$ is the component which pulls the block down the plane and $P\sin\phi$ actually tries to lift the block off the plane, thus reducing the normal force.

The forces may be applied as follows:

forces acting up the plane = forces acting down the plane

$$F_r = P\cos\phi + W\sin\theta$$
$$\mu(W\cos\theta - P\sin\phi) = P\cos\phi + W\sin\theta$$

Rearranging

$$\mu W\cos\theta - \mu P\sin\phi = P\cos\phi + W\sin\theta$$
$$\mu W\cos\theta - W\sin\theta = P\cos\phi + \mu P\sin\phi$$
$$W(\mu\cos\theta - \sin\theta) = P(\cos\phi + \mu\sin\phi) \qquad (6.11)$$

## Example 6.7

A wooden crate is to be pushed up a steel ramp inclined at $20°$ to the horizontal (Figure 6.13). The applied force is inclined at $10°$ to the ramp. If the mass of the crate is 51 kg and the coefficient of friction is 0.37, determine the value of the applied force. Assume the condition of impeding motion.

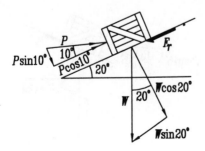

**Fig. 6.13** Crate being pushed up a ramp.

*Solution*

$$W = 51 \times 9.81 = 500\,\text{N}$$

The applied force can be found by using equation (6.10):

$$P \cos \phi = \mu(W \cos \theta + P \sin \phi) + W \sin \theta$$
$$= \mu W \cos \phi + \mu P \sin \phi + W \sin \theta$$
$$P(\cos \phi - \mu \sin \phi) = \mu W \cos \theta + W \sin \theta$$
$$P = \frac{\mu W \cos \theta + W \sin \theta}{\cos \phi - \mu \sin \phi}$$
$$= \frac{0.37 \times 500 \times \cos 20° + 500 \times \sin 20°}{\cos 10° - (0.37 \times \sin 10°)}$$
$$= 374.6\,\text{N}$$

### Example 6.8

The crate in Example 6.7 now has a rope attached and is pulled down the plane (Figure 6.14). If the rope is inclined at 15° to the ramp determine the applied force.

*Solution*

The applied force can be found by using equation (6.11):

$$W(\mu \cos \theta - \sin \theta) = P(\cos \phi + \mu \sin \phi)$$
$$P = \frac{W(\mu \cos \theta - \sin \theta)}{(\cos \phi + \sin \phi)}$$
$$= \frac{500(0.37 \cos 20° - \sin 20°)}{(\cos 15° + 0.37 \sin 15°)}$$
$$= 2.66\,\text{N}$$

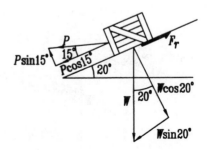

**Fig. 6.14** Crate being pulled down a ramp.

## 6.7  APPLICATION OF FRICTION TO A SCREW THREAD

In whatever shape or form it is found, a screw thread is merely an inclined plane which has been wrapped around a cylinder. Analysis is fairly straightforward but before continuing it is important to become familiar with some screw thread terminology.

### 6.7.1  Single and multi-start threads

A single start thread is where the whole of the thread is part of the same helix as shown in Figure 6.15. There are cases where two or more threads alternate and are known as two or three start threads. Normally single start threads are used where heavy loads are to be applied; however, multi-start threads offer quick transit for only a small angular movement.

Generally, the more starts, the steeper the helix and the faster the transit of the nut. High loads need a flat helix, giving a better mechanical advantage, and are usually a single start thread.

### 6.7.2  Pitch

The distance from a point on one thread to the same point on the next thread measured along the axis of the screw.

### 6.7.3  Lead

The axial distance the nut will travel in one turn of the screw. For a single start thread the lead is the same as the pitch. For a double start thread the lead is twice that of the pitch and this gives

lead = pitch × number of starts                    (6.12)

### 6.7.4  Helix angle

The angle which the thread makes to a plane which is at 90° to the screw axis. From previous work this is the angle of inclination of the inclined plane.

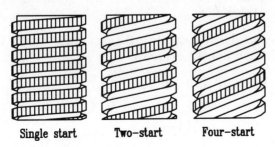

**Fig. 6.15** Single and multi-start threads.

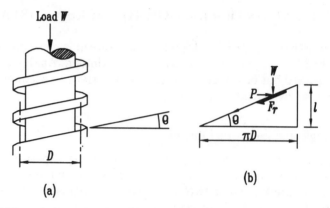

**Fig. 6.16** The relationship of a screw thread to an inclined plane.

Figure 6.16 shows how a screw thread relates to the inclined plane. It can be seen that

$$\tan \theta = \frac{\text{lead}}{\text{circumference}} = \frac{l}{\pi d} \qquad (6.13)$$

where $l$ = lead, $d$ = mean diameter or pitch diameter.

### 6.7.5 The force–torque relationship in turning a screw thread

A torque has to be applied when turning a screw thread, but it should be remembered that a torque is merely a force applied at a radius. If the screw thread is turned by an external force $P$ acting at the mean radius $d/2$, it is equivalent to moving the load up an inclined plane by a horizontal force as shown in Figure 6.16(b).

Equation (6.10) can be used to evaluate $P$:

$$P \cos \phi = \mu(W \cos \theta + P \sin \phi) + W \sin \theta$$

where $\phi$ = angle between applied force and inclined plane,

$\qquad \theta$ = angle of inclination,

$\qquad W$ = axial load on the screw, equivalent to the weight of the block.

It should be noted that $\phi$ and $\theta$ are the same angle since $P$ is horizontal. Equation (6.10) can now be expressed as

$$P = \frac{\mu W \cos \theta + \mu P \sin \theta + W \sin \theta}{\cos \theta}$$

$$= \mu W + \mu P \tan \theta + W \tan \theta$$

The angle of inclination is likely to be different from the angle of friction so this can now be introduced in the form of equation (6.5):

$$\mu = \tan \delta$$

$$P = W \tan \delta + P \tan \delta \tan \theta + W \tan \theta$$

$$P - P \tan \delta \tan \theta = W \tan \delta + W \tan \theta$$

$$P(1 - \tan \delta \tan \theta) = W(\tan \delta + \tan \theta)$$

$$P = \frac{W(\tan \delta + \tan \theta)}{1 - \tan \delta \tan \theta}$$

From trigonometry:

$$\tan(A + B) = \frac{\tan A + \tan B}{1 - \tan A \tan B}$$

gives

$$P = W \tan(\theta + \delta) \qquad (6.14)$$

The torque needed to turn the screw is given by

$$T = P \times \text{mean radius}$$
$$= P \times d/2$$

Inserting equation (6.14) gives

$$T = \frac{d}{2} \times W \tan(\theta + \delta) \qquad (6.15)$$

When $P$ is effectively pulling down the plane, equation (6.15) becomes

$$P = W \tan(\theta - \delta) \qquad (6.16)$$

which gives a torque equation:

$$T = \frac{d}{2} \times W \tan(\theta - \delta) \qquad (6.17)$$

This means that the torque required to lower the load is less than that needed to raise it.

## Example 6.9

The screw-jack shown in Figure 6.17 carries a load of 6 kN and has a square thread, single start screw of 18 mm pitch and 50 mm mean diameter. If the coefficient of friction is 0.22, calculate:

(a) the angle of inclination;
(b) the angle of friction;
(c) the torque to raise the load;
(d) the torque to lower the load.

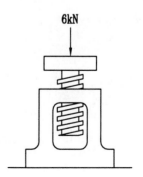

6kN

**Fig. 6.17** Screw-jack.

*Solution*

(a) The angle of inclination can be found by using equation (6.13):

$$\tan\theta = \frac{\text{lead}}{\text{circumference}}$$

$$\theta = \tan^{-1}\left(\frac{0.018}{\pi \times 0.05}\right) = 6.53°$$

(b) The angle of friction can be found using equation (6.5):

$$\mu = \tan\delta$$

or

$$\delta = \tan^{-1}(0.22) = 12.4°$$

(c) The torque required to raise the load can be found using equation (6.15):

$$T = \frac{d}{2} \times W\tan(\theta + \delta)$$

$$= \frac{0.05 \times 6000 \times \tan(6.53° + 12.4°)}{2}$$

$$= 51.47\,\text{N m}$$

(d) The torque to lower the load can be found using equation (6.17):

$$T = \frac{d}{2} \times W\tan(\theta - \delta)$$

$$= \frac{0.05 \times 6000 \times \tan(6.53° - 12.4°)}{2}$$

$$= -15.4\,\text{N m}$$

The negative sign indicates that the load is being lowered.

## Example 6.10

A double start, square-thread screw is used to drive the saddle of a lathe against an axial load of 500 N. The mean diameter of the screw is 55 mm and the pitch is 6 mm. If the coefficient of friction is 0.18, find:

(a) the angle of inclination;
(b) the angle of friction;
(c) the torque needed to rotate the screw.

*Solution*

(a) The angle of inclination can be found using equation (6.12):

$$\text{lead} = \text{pitch} \times \text{number of starts}$$

and equation (6.13):

$$\tan \theta = \frac{\text{lead}}{\text{circumference}}$$

$$\theta = \tan^{-1}\left(\frac{2 \times 0.006}{\pi \times 0.055}\right) = 4.0°$$

(b) The angle of friction can be found using equation (6.5)

$$\mu = \tan \delta$$

or

$$\delta = \tan^{-1}(0.18) = 10.2°$$

(c) The torque needed to rotate the screw can be found using equation (6.15):

$$T = \frac{d}{2} \times W \tan(\theta + \delta)$$

$$= \frac{0.055 \times 500 \times \tan(4.0° + 10.2°)}{2}$$

$$= 3.48 \, \text{N m}$$

## 6.8 SUMMARY

$\mu$ = coefficient of friction
$F_r$ = friction force (resistance to motion)
$N$ = normal force (force at 90° to friction plane)
$P$ = externally applied force
$\theta$ = angle of inclined plane
$\phi$ = angle between applied force and inclined plane
$\delta$ = angle of friction

For impending motion:

$$P = F_r = \mu N \tag{6.3}$$

Angle of friction:

$$\mu = \tan \delta \tag{6.5}$$

External force pulling up and parallel to an inclined plane:

$$P = W \sin \theta + \mu W \cos \theta \tag{6.7}$$

Oblique external force pushing up and into an inclined plane:

$$P \cos \phi = \mu(W \cos \theta + P \sin \phi) + W \sin \theta \tag{6.10}$$

Oblique external force pulling down and out of inclined plane:

$$W(\mu \cos \theta - \sin \theta) = P(\cos \phi + \mu \sin \phi) \tag{6.11}$$

### 6.8.1 Applications to screws

Applied force up the plane:

$$P = W \tan(\theta + \delta) \tag{6.14}$$

Torque up the plane:

$$T = \frac{Wd}{2} \tan(\theta + \delta) \tag{6.15}$$

Applied force down the plane:

$$P = W \tan(\theta - \delta) \tag{6.16}$$

Torque down the plane:

$$T = \frac{Wd}{2} \tan(\theta - \delta) \tag{6.17}$$

### 6.9  PROBLEMS

1. A body having a mass of 40 kg rests on a horizontal table and it is found that the smallest horizontal force required to move the body is 90 N. Find the coefficient of friction.
2. A chest of drawers has a mass of 41 kg and rests on a tile floor whose coefficient of friction is 0.25. Find the smallest horizontal force P needed to move the chest if a man pushes it across the floor. If the man has a mass of 70 kg, find the lowest coefficient of friction between his shoes and the floor so that he does not slip.
3. A block of stone is hauled along a horizontal floor by a force applied through a rope inclined at 20° to the horizontal. If the block has a mass of 40 kg and the coefficient of friction between the stone and the floor is 0.3, find the tension in the rope.
4. A body of mass 250 kg rests on a horizontal plane. Determine the external force needed to move the body if the coefficient of friction is 0.4 and the force acts:
   (a) in an upward direction at 20° to the plane; and
   (b) in a downward direction at 10° to the plane.
5. A crate just begins to slide down a rough plane inclined at 25° to the horizontal. If the mass of the crate and its contents is 500 kg, find the horizontal force required to haul it up the plane with uniform velocity.
6. A body of weight 500 N is prevented from sliding down a plane inclined at 20° to the horizontal by the application of a force of 50 N acting upwards and parallel to the plane. Find the coefficient of friction. The body is now pulled up the plane by a force acting at 30° to the horizontal. Find the value of the force.
7. Find the torque to raise a load of 6000 N by a screw-jack having a double start square thread, with two threads per centimetre and a mean diameter of 60 mm. Take the coefficient of friction to be 0.12.

8. A nut on a single start square thread is locked up against a shoulder by a torque of 6 N m. If the thread pitch is 5 mm and the mean diameter is 60 mm, determine:
   (a) the axial load on the screw in kg;
   (b) the torque required to loosen the nut.
   Take the coefficient of friction to be 0.1.

# Introduction to vibrations and simple harmonic motion

7

## 7.1  AIMS

- To introduce the basic concepts of vibration.
- To define the terms which describe the theory of vibration.
- To explain the causes of vibration in a system.
- To explain the analysis of vibration in simple systems.
- To introduce simple harmonic motion, the most basic theoretical form of vibration.

## 7.2  INTRODUCTION TO VIBRATION

The study of vibration is concerned with the oscillation of masses and the associated forces. Oscillations occur because of the way masses and the elasticity of materials combine. When vibrating, the mass of a material possesses momentum which tries to stretch and squash the material. This is analogous to a mass suspended on a spring.

All bodies possess mass and in so doing also possess a degree of elasticity. It can be said that everything will vibrate to some extent. For example, hold up a house brick with one hand and strike the brick with a hammer. A short 'ding' can be heard. The vibration is at a fairly high frequency and is perceived as sound, but dies away quickly. Hold a ruler over the edge of a desk and flick the end. The oscillations are so slow that they can be seen. Both the brick and the ruler oscillate because there is energy within the mass, trying to move it. The material itself, however, provides the 'springiness' which allows the mass to vibrate.

There are generally two classes of vibration, free and forced vibration.

### 7.2.1  Free vibration

This takes place when energy within the mass–elastic system creates internal forces which push and pull the mass into oscillatory motion. A system which is allowed to vibrate in this natural way will do so at one of its natural frequencies. A **natural frequency** is a particular 'speed' of oscillation and is a product of the amount of mass and the elastic distribution within the vibrating system. It is the frequency at which the

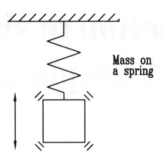

**Fig. 7.1** Mass on a spring – free vibration.

system naturally prefers to vibrate. Examples of free vibrating systems were considered earlier with the house brick and the ruler. An analogous system is shown in Figure 7.1 as a mass on a spring.

### 7.2.2  Forced vibration

This takes place when forces from outside the mass–elastic system push and pull the mass. Often these external forces are themselves oscillating and cause the system to oscillate at the same speed. An excellent example of forced vibration is when the fuel ignites within an internal combustion engine, and the subsequent oscillations of the engine impart forces to the vehicle chassis. These are perceived as vibration and engine noise within the passenger compartment.

Another good example is that of a washing machine. When set to spin the washing may congregate in lumps within the drum. As the drum spins at high speed the mass of washing imparts a force on the drum, due to centrifugal force, which rotates as the drum rotates. This rotating force is passed to the washing machine body as an oscillating force. The response of the washing machine is to oscillate at the same speed, sometimes so violently as to 'walk' across the floor. Figure 7.2 illustrates this example.

**Fig. 7.2** Forced vibration in a washing machine.

Any mass–elastic system possesses natural frequencies and if an oscillating force is applied at the natural frequency, then the system will oscillate violently and can be said to resonate. In a resonating structure, such large and dangerous movements may occur that the structure will disintegrate. Structures such as bridges, buildings and aeroplane wings have been known to break up when resonant frequencies have been encountered. It is, therefore, of major importance that natural frequencies are able to be predicted. Section 7.5 describes how this may be achieved for simple systems.

Most vibrating systems are subject to varying degrees of damping. This is the term given to energy dissipation by friction and other resistances. Damping tends to slow the speed of oscillation slightly, though small damping values will have little effect on the natural frequency. Large damping values are of great importance, however, because they limit the range of oscillation (amplitude) at resonance.

The implications of damping have been omitted from this text in the interests of introducing the concepts of vibration in an uncomplicated manner.

## 7.3 HARMONIC MOTION

Oscillations may repeat themselves at regular intervals as in the case of a clock pendulum or they may display a great deal of irregularity as in earth tremors.

The principal concern here is in describing **periodic motion**, which is motion repeated in regular intervals of time. The time taken to complete one oscillation is called the **period** and is given the symbol $\tau$. If $\tau$ is the time taken for one cycle, then its reciprocal $1/\tau$ gives the number of cycles completed per second, and is termed the **frequency**, i.e.

$$f = \frac{1}{\tau} \tag{7.1}$$

where $f$ = frequency (cycles/s or hertz),
$\tau$ = the period (s).

The simplest form of periodic motion is called **harmonic motion** and is demonstrated by a mass suspended on a spring, oscillating with an up-and-down motion. If the amplitude of this motion is plotted against time, a sine wave is produced as shown in Figure 7.3.

It is necessary to be able to predict the motion which can be achieved by utilizing the equation which describes the sine wave. At any instant in time it is possible to determine the height of the sine wave from the equation:

$$x = A \sin \omega t \tag{7.2}$$

where $x$ = height of the sine wave at any instant in time (m),
$A$ = maximum height of the sine wave (m),
$\omega$ = frequency (rad/s),
$t$ = elapsed time (s).

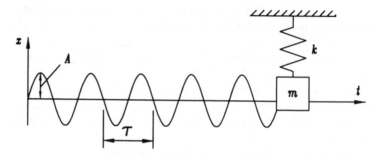

**Fig. 7.3** Sine wave as a record of harmonic motion.

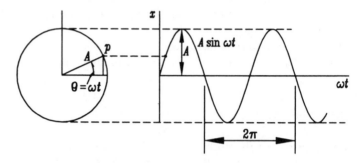

**Fig. 7.4** Harmonic motion related to a circular path.

This equation can be explained by relating a single cycle of the sine wave to a circle, as shown in Figure 7.4:

- The amplitude $A$ is the maximum height of the sine wave and is also the radius of the circle.
- The height of the sine wave $x$ at any time $t$ can be related to the circle and a triangle can be drawn as shown in Figure 7.4.
- The hypotenuse of the triangle is the **amplitude** $A$.
- To find the amplitude $x$ a simple trigonometric sine ratio can be used, i.e.

$$x = A \sin \theta \tag{7.3}$$

The angle $\theta$ will vary with time as the sine wave completes its cycle and, therefore, needs to be related to time.

The velocity of a point on the sine wave may be related to the angular velocity of a similar point as it traces out the circle.

Since the angular velocity $\omega$ is measured in rad/s, it may be multiplied by the elapsed time $t$ to convert to an angle $\theta$ in rad:

$$\theta = \omega t \quad \text{or in units} \quad \frac{\text{rad}}{\cancel{s}} \times \cancel{s}$$

Equation (7.3) can now be written as

$$x = A \sin \omega t$$

which is recognizable as equation (7.2).

At any point in the cycle, the height $x$ may be found by inserting the frequency $\omega$ and the elapsed time $t$.

Before leaving this discussion it should be noted that frequency has been mentioned in terms of $f$ cycles per second (hertz) and $\omega$ rad/s. Both quantities are valid expressions; however, they may be used in different ways. Frequency expressed in cycles/s is more easily visualized, especially since slow oscillations may sometimes be counted. Frequency expressed in rad/s is used principally in calculation. The conversion between the two, also incorporating periodic time $\tau$, is as follows:

$$f = \frac{\omega}{2\pi} = \frac{1}{\tau} \tag{7.4}$$

**Example 7.1**

A body is suspended on a spring and oscillates at a frequency of 2 Hz (Figure 7.5). If the maximum displacement amplitude $A$ is measured as 10 mm, determine:

(a) the frequency $\omega$ in rad/s;
(b) the displacement $x$ after an elapsed time $t$ of 0.07 s.

*Solution*

(a) The frequency in rad can be found by using equation (7.4):

$$f = \frac{\omega}{2\pi}$$

Now $f = 2$ Hz so that

$$\omega = 2\pi f = 2\pi \times 2 = 12.56 \text{ rad/s}$$

(b) The displacement can be found by using equation (7.2):

$$x = A \sin \omega t$$

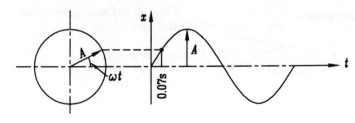

**Fig. 7.5**

Now $A = 10$ mm or $0.01$ m, $t = 0.07$ s, $\omega = 12.56$ rad/s, so that

$x = 0.01 \times \sin(12.56 \times 0.07) = 7.7$ mm

## 7.4 SIMPLE HARMONIC MOTION

Simple harmonic motion, or SHM, is defined as the periodic motion of a body, or point, the acceleration of which is:

1. always towards a fixed point lying on its path;
2. proportional to its displacement from that point.

In order to visualize this definition consider a pendulum as shown in Figure 7.6. When plotted against time the motion is that of a sine wave and behaves with SHM.

When the pendulum bob is placed at point a, the action of gravity imparts maximum acceleration towards the centre point O. As the pendulum passes point O it hangs vertically and cannot be accelerated further, i.e. acceleration is zero. On the climb up to point b the pendulum is slowed by the action of gravity. It therefore possesses negative acceleration, and the true acceleration can still be taken as acting towards the centre point O. Thus part (a) of the definition is fulfilled.

The values of acceleration at points a and b are equal in magnitude, but more importantly, they represent a maximum which corresponds to the maximum amplitude $A$ on the sine wave.

As the pendulum moves towards the centre point O the value of the acceleration decreases and is proportionally dependent upon the dis-

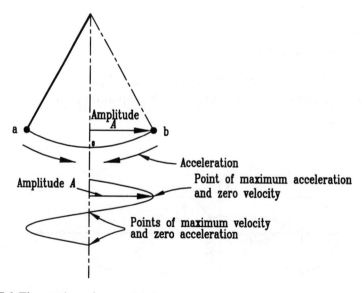

**Fig. 7.6** The motion of a pendulum.

placement from point O. At the instant the pendulum passes through point O there is no acceleration, but at this point it may be seen that the velocity is a maximum.

## 7.5 CALCULATION OF NATURAL FREQUENCY

The preceding discussion has described the importance of predicting natural frequency but this needs to be done in a way which takes account of the physical aspects of the vibrating system. For a simple mass–elastic system there would be mass $M$ and spring stiffness $k$.

### 7.5.1 Spring stiffness

Spring stiffness, designated $k$, is the 'strength' of the spring. Due to high loads the spring stiffness of a vehicle suspension will have a high value of $k$, while the spring in a ball-point pen will have a low value of $k$.

The value of $k$ is the force required to stretch or compress a spring over a certain distance. Double the force and the distance will also be doubled. For a uniform open coiled helical spring this relationship is largely linear, as shown in Figure 7.7. The value $k$ for any spring or elastic component can be determined by considering:

$$k = \frac{\text{force}}{\text{extension}} \quad \text{with units of N/m}$$

The spring stiffness $k$ is actually the slope of the line and can be shown by hanging masses on, for example, an elastic band.

### 7.5.2 Calculation of displacement, velocity and acceleration

During an oscillation the mass can be seen to: accelerate, slow down, stop, change direction, accelerate and slow down twice every cycle. The mass must therefore:

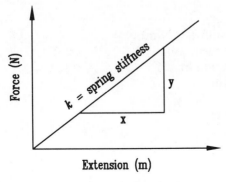

Fig. 7.7 Force extension graph for a typical spring.

- be displaced (m);
- possess varying values of velocity (m/s);
- possess varying values of acceleration ($m/s^2$).

These changes within the dynamic system were discussed using the example of the pendulum, but the values of velocity and acceleration may be calculated using the displacement equation (7.2).

Displacement:

$$x = A \sin \omega t \tag{7.2}$$

Differentiating with respect to time gives velocity:

$$\dot{x} = \omega A \cos \omega t \tag{7.5}$$

Differentiating with respect to time gives acceleration:

$$\ddot{x} = -\omega^2 A \sin \omega t \tag{7.6}$$

Velocity components are used when damping is present and since a basic mass–elastic system does not possess damping, equation (7.5) may be ignored at this stage.

It follows that the maximum velocity and acceleration are given by

$$\dot{x}_{max} = \omega A \quad \text{(when displacement is zero)} \tag{7.7}$$

$$\ddot{x}_{max} = -\omega^2 A \quad \text{(when displacement is } A) \tag{7.8}$$

Mass and elasticity parameters and the displacement and acceleration equations can now be combined as in Newton's second law:

$$F = ma$$

This equation can now be written in terms which are used to describe oscillation, so that for equilibrium

$$m\ddot{x} - kx = 0 \tag{7.9}$$

where $k$ = spring stiffness (N/m),
$\quad x$ = displacement (m),
$\quad \ddot{x}$ = acceleration ($m/s^2$),
$\quad m$ = mass (kg).

Transposing gives

$$\frac{k}{m} = -\frac{\ddot{x}}{x}$$

Inserting equations (7.2) for displacement and (7.6) for acceleration

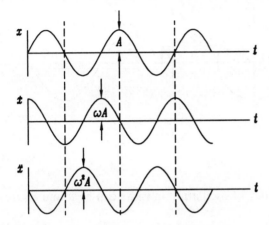

**Fig. 7.8** Plots of displacement, velocity and acceleration.

gives

$$\frac{k}{m} = -\frac{\ddot{x}}{x} = -\frac{\omega^2 A \sin \omega t}{A \sin \omega t}$$

or

$$\omega^2 = \frac{k}{m} = -\frac{\ddot{x}}{x} \qquad (7.10)$$

giving

$$\omega = \sqrt{\frac{k}{m}} \qquad (7.11)$$

and

$$\omega = \sqrt{-\frac{\ddot{x}}{x}} \qquad (7.12)$$

where $\omega =$ natural frequency (rad/s).

These equations show how the natural frequency may be calculated from the physical attributes of a vibrating system. Figure 7.8 shows a plot of equations (7.2), (7.5) and (7.6) which reveals that they are all sinusoidal, harmonic motion. It should be noted that velocity lags displacement by 90° or $\pi/2$ rad and acceleration lags displacement by 180° or $\pi$ rad.

## Example 7.2

A body which moves with simple harmonic motion has a maximum acceleration of 5 m/s² (Figure 7.9). If the maximum amplitude is 30 mm, find the frequency of oscillation:

(a) in rad/s
(b) in Hz.

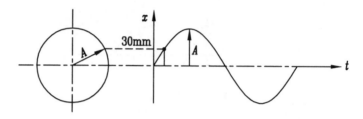

**Fig. 7.9**

*Solution*

(a) The frequency of oscillation in rad/s can be found by using equation (7.12):

$$\omega = \sqrt{-\frac{\ddot{x}}{x}}$$

Now $\ddot{x} = -5\,\text{m/s}^2$, $x = 30\,\text{mm}$ or $0.03\,\text{m}$, so that

$$\omega = \sqrt{\frac{5}{0.03}} = 12.9\,\text{rad/s}$$

(b) The frequency of oscillation in Hz can be found by using equation (7.4):

$$f = \frac{\omega}{2\pi}$$

Now $\omega = 12.9\,\text{rad/s}$, so that

$$f = \frac{12.9}{2\pi} = 2.05\,\text{Hz}$$

**Example 7.3**

A block of mass 10 kg vibrates with SHM at a frequency of 100 Hz (Figure 7.10). The total displacement of the mass has been measured at 4 mm. Determine:

(a) the elapsed time when the mass is 0.5 mm from the extremity of its stroke;
(b) the velocity of the mass when it is 0.5 mm from the extremity of its stroke;
(c) the acceleration of the mass when it is 0.5 mm from the extremity of its stroke;
(d) the maximum force acting on the block.

*Solution*

(a) The elapsed time can be calculated by using equation (7.2):

$$x = A \sin \omega t$$

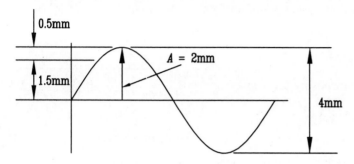

**Fig. 7.10**

Transposing for $t$ gives

$$\omega t = \sin^{-1}\left(\frac{x}{A}\right)$$

$$t = \frac{1}{\omega}\sin^{-1}\left(\frac{x}{A}\right)$$

Now

$$A = 2.0\,\text{mm} \quad \text{or} \quad 0.002\,\text{m}$$

$$x = 2.0 - 0.5 = 1.5\,\text{mm} \quad \text{or} \quad 0.0015\,\text{m}$$

$$\omega = f \times 2\pi = 100 \times 2\pi = 628.3\,\text{rad/s}$$

so that

$$t = \frac{1}{628.3}\sin^{-1}\left(\frac{0.0015}{0.002}\right) = 1.35 \times 10^{-3}\,\text{s}$$

(b) The velocity of the mass can be found by using equation (7.5):

$$\dot{x} = \omega A \cos \omega t$$

Now $\omega = 628.3\,\text{rad/s}$, $A = 0.002\,\text{m}$, $t = 1.35 \times 10^{-3}\,\text{s}$, so that

$$\dot{x} = 628.3 \times 0.002 \times \cos(628.3 \times 1.35 \times 10^{-3})$$

$$= 0.831\,\text{m/s}$$

(c) The acceleration of the mass can be found using equation (7.6):

$$\ddot{x} = -\omega^2 A \sin \omega t$$

Now $\omega = 628.3\,\text{rad/s}$, $A = 0.002\,\text{m}$, $t = 1.35 \times 10^{-3}\,\text{s}$, so that

$$\ddot{x} = -628.3^2 \times 0.002 \times \sin(628.3 \times 1.35 \times 10^{-3})$$

$$= 592.2\,\text{m/s}^2$$

Alternatively equation (7.10) can be used:

$$\omega = \sqrt{-\frac{\ddot{x}}{x}}$$

Transposing for $\ddot{x}$ gives

$$\ddot{x} = \omega^2 x$$

Now $\omega = 628.3\,\text{rad/s}$ and $x = 0.0015\,\text{m}$, so that

$$\ddot{x} = 628.3^2 \times 0.0015 = 592.2\,\text{m/s}^2$$

(d) The force acting on the mass can be found by using Newton's second law in the form of equation (3.2). Maximum acceleration is

$$\ddot{x}_{max} = 628.3^2 \times 0.002 = 789.5\,\text{m/s}^2$$

$$F = ma$$

Now $m = 10\,\text{kg}$ so that

$$F = 10 \times 789.5 = 7.895\,\text{kN}$$

## 7.6 GLOSSARY OF TERMS

**Amplitude** ($A$)   The maximum height of an oscillation measured from the axis of the sine wave to the top of the peak. Often amplitude refers to displacement but it may also refer to values of velocity, acceleration or force.

Note that the total peak-to-peak distance of a sine wave is twice the amplitude.

**Damping**   The term given to energy dissipation by friction and other resistances. Its most obvious effect is to reduce the value of displacement by absorbing the energy which causes the oscillation.

**Dynamic system**   A collection of connected components which continue to move relative to each other.

**Frequency**   In terms of oscillation it is a measure of how often a cycle repeats itself per unit of time. Frequency is measured either in cycles per second (designated hertz or Hz) or in radians per second.

Frequency in cycles per second, designated '$f$', is easy to visualize and is used as a comparison of oscillation speed. Frequency measured in radians per second, designated '$\omega$', is used principally in calculation.

**Harmonic motion**   The simplest form of periodic motion. It can also be described as motion which describes a sinusoidal path during its oscillation.

**Mass–elastic system** An oscillating system in which there is no damping. It comprises an element of mass and a spring element. It should be noted that few systems in practice can be reduced to such an idealistic level.

**Natural frequency** For a particular combination of mass and elasticity, this is a frequency at which the system prefers to vibrate. The lowest of these frequencies is termed the **fundamental natural frequency** and is the frequency at which the lowest resonance occurs. It is also termed 'the first harmonic'. For any given system there usually occur higher natural frequencies which are termed second harmonic, third harmonic, etc.

**Oscillation** The movement of a mass, to and fro, between two extreme points.

**Period** The time taken to complete one cycle of oscillatory motion, designated $\tau$ and measured in seconds.

**Periodic motion** Any motion which is repeated, where the repeat time is exactly the same, i.e. the period $\tau$. The motion may vary from sinusoidal to being very complex, perhaps supporting numerous different but combined frequencies.

**Resonate/resonance** A condition of an oscillating system where an excitation frequency coincides with a natural frequency. The results of such a combination may vary from uncomfortable vibrations to violent oscillation which is self-destructive.

**Vibration** (see also oscillation) Motion to and fro, especially of parts of a fluid or an elastic solid whose equilibrium has been disturbed. Vibration and oscillation share the same definition.

## 7.7 SUMMARY

Displacement:

$$x = A \sin \omega t \quad \text{(m)} \tag{7.2}$$

Velocity:

$$\dot{x} = \omega A \cos \omega t \quad \text{(m/s)} \tag{7.5}$$

Acceleration:

$$\ddot{x} = -\omega^2 A \sin \omega t \quad \text{(m/s}^2\text{)} \tag{7.6}$$

Maximum velocity:

$$\dot{x}_{\text{max}} = \omega A \quad \text{(m/s)} \tag{7.7}$$

Maximum acceleration:

$$\ddot{x}_{\text{max}} = -\omega^2 A \quad \text{(m/s}^2\text{)} \tag{7.8}$$

where $A$ = maximum amplitude (m),
    $\omega$ = frequency   (rad/s),
    $t$ = elapsed time (s),
    $\omega t$ = angle $\theta$ (rad).
    $f$ = frequency (cycles/second) or (hertz, designated Hz)
    $\tau$ = period (s)

$$f = \frac{1}{\tau} \tag{7.1}$$

$$f = \frac{\omega}{2\pi} \tag{7.4}$$

$$\omega = \sqrt{\frac{k}{m}} \tag{7.11}$$

$$\omega = \sqrt{-\frac{\ddot{x}}{x}} \tag{7.12}$$

where $k$ = spring stiffness (N/m),
    $m$ = mass (kg).

## 7.8   PROBLEMS

*Advice:* Put the calculator in radian mode before trying these problems.

1. A certain harmonic motion has an amplitude of 0.7 m and a period of 1.3 s. Find:
   (a) the frequency of oscillation;
   (b) the maximum velocity;
   (c) the maximum acceleration.
2. Vibration measuring instruments indicate that a body vibrates with SHM at a frequency of 10 Hz, with a maximum acceleration of 8 m/s². Find the displacement amplitude.
3. A point in a mechanism moves with SHM, with an amplitude of 0.4 m and a frequency of 2 Hz. Find:
   (a) the maximum acceleration;
   (b) the velocity when the point is 0.15 m from the central position of vibration.
4. An instrument of mass 20 kg oscillates with SHM of amplitude 20 mm. When the instrument is 20 mm from the centre of oscillation the force acting on it is 20 N. Find the frequency of oscillation.
5. A 2 kg mass moving with SHM has a period of 0.7 s and an amplitude of 200 mm. Find the maximum kinetic energy of the mass. Use $KE = \frac{1}{2}mv^2$.
6. The valve in an internal combustion engine moves vertically with SHM due to the action of the cam. The full lift, i.e. twice the amplitude,

is 12 mm and is completed in 0.4 s. When the valve has lifted 2.5 mm from its seat (lowest position), find:
(a) its velocity;
(b) its acceleration.

7. A cam causes a follower of mass 4 kg to oscillate with SHM. The follower rises to a maximum lift of 40 mm during one-half of the cam revolution and returns during the second half of the revolution. If the cam rotates at 300 rev/min find:
(a) the maximum force on the follower;
(b) the kinetic energy of the follower when 15 mm from the extremity of its travel.

# Stresses and strains $\boxed{8}$

## 8.1 AIMS

- To discuss the stresses and strains in components resulting from direct loading.
- To introduce the concept of a factor of safety to ensure that working stresses are within acceptable limits.
- To analyse the stresses in thin cylinders.
- To discuss the lateral strain resulting from direct loading and to introduce Poisson's ratio.
- To discuss the stresses and strains in components resulting from shear loading.

## 8.2 BEHAVIOUR OF MATERIALS

If a bar of a material is subjected to an axial force, there will be a change in its length. This is true for all materials, but the actual change in the length will vary for different materials.

Consider a vertical bar supported at the top and subjected to an axial force, as shown in Figure 8.1. As a result of the force there will be an equal and opposite reaction in the support at the top, so the bar can be viewed as being subjected to two external forces at either end that are stretching the material. The effect on the bar is to cause its length to increase, as shown in Figure 8.1. The bar is said to be subjected to a 'tensile' force and, as a consequence, the bar is in **tension**.

With loading in the opposite direction, the axial forces cause a reduction in length and the bar is said to be subjected to a 'compressive' force, so that the bar is in **compression**.

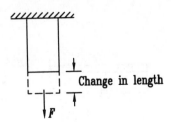

**Fig. 8.1** Extension of a bar.

If the bar is stretched sufficiently it will eventually break. The extent of the change of length determines the characteristics of the material. If a bar is subjected to a tensile force that causes a large increase in length before it breaks, the material is said to be **ductile**. Alternatively, if the bar breaks suddenly with a very short extension, the material is said to be **brittle**.

In terms of common everyday materials, paper is a reasonably good ductile material whereas glass is a brittle material and will shatter quite readily. Ductile materials cope well with tensile loads whereas brittle materials are better with compressive loads. The tall columns in a medieval cathedral are built of stone, which is a brittle material. The columns are stable and have survived the centuries because they are subjected to compressive forces, the forces being due to the weight of the roof and the respective reactions from the ground. If the columns were suddenly subjected to tensile forces they would fail.

In terms of engineering materials steel and aluminium alloys have good ductile properties whereas cast iron is a brittle material. What must be appreciated is that all materials break if subjected to a large enough force. It is not the force directly that breaks the material but the extension resulting from the force.

Although it is easier to visualize the failure of a bar in tension, failure can also occur in compression, if the force is great enough. With a compressive failure, the material may crumble apart. Alternatively, a bar subjected to a compressive force may suddenly buckle and bend out of shape. Because there are several possible modes of failure in compression, the topic is outside the scope of this book.

The behaviour of materials under tensile loads depends on the size of the bar and the magnitude of the applied forces. The change of length resulting from a given force depends on the cross-sectional area and original length. These can be related in terms of the 'stress' and 'strain' of a bar or component.

## 8.3 DIRECT LOADING

A bar subjected to the sort of loading illustrated in Figure 8.1, in which the applied forces are parallel to the longitudinal axis, is being acted upon by **direct** forces.

Consider the rectangular bar shown in Figure 8.2. The bar is represented by a free-body diagram with an external applied force of $F$. As a result

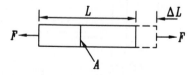

**Fig. 8.2** A bar subjected to direct loading.

of the applied force, there will be a change of length of the bar, denoted by $\Delta L$. Clearly, the change in length depends on the original length of the bar. A bar 2 m in length will stretch twice as far as a similar bar 1 m long. The change in length is also dependent on the cross-sectional area on which the force is acting.

### 8.3.1 Direct stress and strain

A force results in a **stress** in the material. If the applied force is $F$ and the cross-sectional area is $A$, the force can be considered to act uniformly over the whole of the area $A$. This uniform load distribution is called the stress, where

$$\text{stress} = \frac{\text{force}}{\text{cross-sectional area}}$$

In terms of symbols, the direct stress is designated as $\sigma$ where

$$\sigma = \frac{F}{A} \tag{8.1}$$

with units of $(\text{N/m}^2)$ or Pa. This is valid irrespective of whether the bar or component is in tension or compression.

It is the magnitude of the stress that determines the change in length of the bar, together with the original length. It is possible to relate the change of length to the original length by means of the **strain**. In terms of a change of length, $\Delta L$, relating to an original length, $L$, the strain, $\varepsilon$, can be defined as

$$\varepsilon = \frac{\Delta L}{L} \tag{8.2}$$

Since this represents a length divided by a length, $\varepsilon$ is clearly dimensionless, having no units.

### Example 8.1

A stepped bar has a maximum diameter of 50 mm and a minimum diameter of 25 mm (Figure 8.3). If a tensile load of 50 kN is applied, calculate the direct stress in each section.

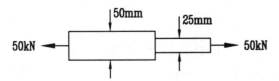

**Fig. 8.3**

*Solution*

The stress can be found using equation (8.1):

$$\sigma = \frac{F}{A}$$

For the large diameter:

$$A_1 = \frac{\pi}{4}d_1^2 = \frac{\pi}{4} \times \frac{50^2}{10^6} = 1.96 \times 10^{-3}\,\text{m}^2$$

and

$$\sigma_1 = \frac{50 \times 10^3}{1.96 \times 10^{-3}} = 25.5 \times 10^6\,\text{Pa}$$

$$= 25.5\,\text{MPa}.$$

For the small diameter:

$$A_2 = \frac{\pi}{4} \times \frac{25^2}{10^6} = 4.91 \times 10^{-4}\,\text{m}^2$$

and

$$\sigma_2 = \frac{50 \times 10^3}{4.91 \times 10^{-4}} = 101.8 \times 10^6\,\text{Pa}$$

$$= 101.8\,\text{MPa}$$

**Example 8.2**

A cylinder has a cover plate attached by eight bolts of 12 mm diameter (Figure 8.4). If the inside diameter of the cylinder is 250 mm, what pressure will the cylinder withstand if the maximum allowable stress in the bolts is 40 MPa?

*Solution*

The total load on the bolts can be found from equation (8.1):

$$\sigma = \frac{F}{A} \quad \text{so that} \quad F = \sigma \times A$$

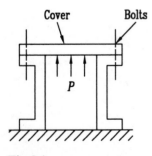

Cover    Bolts

*P*

**Fig. 8.4**

For each bolt:

$$A = \frac{\pi}{4}d^2 = \frac{\pi}{4} \times \frac{12^2}{10^6} = 1.131 \times 10^{-4}\,\text{m}^2$$

For 12 bolts:

$$\text{total area} = 12 \times 1.131 \times 10^{-4} = 1.357 \times 10^{-3}\,\text{m}^2$$

Total load due to the pressure is

$$F_T = 40 \times 10^6 \times 1.357 \times 10^{-3} = 54\,288\,\text{N}$$

and pressure is

$$P = \frac{F_T}{\text{area of cylinder}}$$

$$= \frac{54\,288}{\pi/4 \times (250)^2/10^6} = 1.106 \times 10^6 \,\text{Pa}$$

$$= 1.106 \,\text{MPa}.$$

### 8.3.2 Stress–strain curve

If sample bars of a material are tested under laboratory conditions so that corresponding values of stress and strain can be recorded, the resulting curve illustrates the behaviour of the material.

Figure 8.5 shows a typical stress–strain curve for a low carbon steel, generally referred to as a 'mild' steel. Steel is an alloy of iron and carbon. In the case of mild steel, the carbon represents only 0.1–0.2% of the total alloy.

The curve shown in Figure 8.5 has two distinct regions:

1. elastic region, ab;
2. plastic region, be.

This is typical of a ductile material, which is characterized by a large plastic extension before failure.

In the elastic region, ab, an increase in stress results in a proportional increase in strain. If all the load is removed, so that the stress is zero, the bar returns to its original length.

If the loading is increased beyond the elastic region, the bar starts to be permanently deformed. Within the plastic region, the strain is no longer proportional to the stress and a large increase in strain can result from a small increase in stress. The region of the curve bc is peculiar to the behaviour of mild steel. It represents 'perfect plastic' behaviour in which the strain increases with no increase in stress, but this only occurs for a short extension. From c to d, strain hardening causes a change to the crystalline structure of the steel that results in an increase in stress. From d to e, the bar or specimen actually undergoes a reduction in cross-sectional area due to 'necking' (Figure 8.6), until failure occurs at point e.

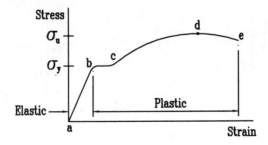

**Fig. 8.5** Stress–strain curve for mild steel.

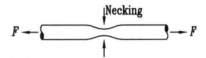

**Fig. 8.6** Necking of a bar in tension.

At the point where the change from elastic to plastic behaviour takes place, there is a yielding of the material and the stress, $\sigma_y$, is referred to as the yield stress. The maximum stress, achieved at point d, is referred to as the ultimate tensile stress, $\sigma_u$.

Other ductile materials have a stress–strain curve as shown in Figure 8.7(a) and, for comparison, a typical stress–strain curve for a brittle material is shown in Figure 8.7(b). The brittle material exhibits a negligible amount of plastic deformation before failure, while the ductile material is characterized by large plastic deformation.

Properties, including values of $\sigma_y$ and $\sigma_u$, are quoted for a range of metals in Appendix A.

### 8.3.3 Elastic relationship

Within the elastic region the behaviour of the material is characterized by the strain being proportional to the stress. This can be expressed as

$$\sigma \propto \varepsilon$$

or, alternatively,

$$\frac{\sigma}{\varepsilon} = \text{constant}$$

The value of the constant depends on the particular material, but is given the symbol $E$:

$$E = \frac{\sigma}{\varepsilon} \tag{8.3}$$

where $E$ is termed the **modulus of elasticity** with units of Pa.

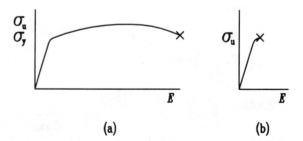

(a)        (b)

**Fig. 8.7** Typical stress–strain curves for ductile and brittle materials.

Equation (8.3) is sometimes referred to as Hooke's law. Although the concepts of stress and strain were not formally defined until the early part of the nineteenth century, the realization that the extension is proportional to the applied force was explained by Robert Hooke as early as 1676.

The modulus of elasticity, $E$, is a measure of the **stiffness** of the material. The greater the strain for a given stress, the smaller the value of $E$ and the lower the stiffness. Alternatively, the smaller the strain for a given stress, the greater the value of $E$ and the greater the stiffness. Values of $E$ vary from 7 GPa for some soft woods through to over 1000 GPa for diamond.

Typical values of $E$ for various metals are given in Appendix A.

**Example 8.3**

A steel bar of 25 mm diameter is subjected to a tensile force of 8 kN (Figure 8.8). If the bar is originally 1.2 m long, determine the increase in length. Take $E$ for steel as 200 GPa.

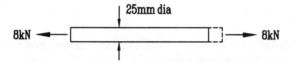

**Fig. 8.8**

*Solution*

The stress can be found using equation (8.1):

$$\sigma = \frac{F}{A}$$

For the bar:

$$A = \frac{\pi}{4} d^2 = \frac{\pi}{4} \times \frac{25^2}{10^6} = 4.91 \times 10^{-4} \, \text{m}^2$$

and

$$\sigma = \frac{8 \times 10^3}{4.91 \times 10^{-4}} = 16.3 \times 10^6 \, \text{Pa}$$

$$= 16.3 \, \text{MPa}$$

The strain can be found using equation (8.3):

$$E = \frac{\sigma}{\varepsilon}$$

or

$$\varepsilon = \frac{\sigma}{E} = \frac{16.3 \times 10^6}{200 \times 10^9} = 8.15 \times 10^{-5}$$

The increase in length can be found using equation (8.2):

$$\varepsilon = \frac{\Delta L}{L}$$

and

$$\Delta L = \varepsilon \times L$$
$$= 8.15 \times 10^{-5} \times 1.2 = 9.78 \times 10^{-5}\,\text{m}$$
$$0.0978\,\text{mm}$$

## 8.4 FACTOR OF SAFETY

In Example 8.3 the stress was 16.3 MPa for a load of 8 kN. If the ultimate tensile stress for the steel was 400 MPa, the load could have been increased to

$$8 \times \frac{400}{16.3} = 196\,\text{kN}$$

before the bar failed. The ratio of the ultimate tensile stress to the actual stress represents a 'factor of safety' in this case.

Factors of safety are necessary to take into account uncertainties regarding the stress analysis for a particular situation. The uncertainties can relate to either the loading of the component under consideration or the material properties.

Material properties are very significant. The stress values quoted in Appendix A are just typical values; they do not represent exact values. Therefore, the ultimate tensile stress of several samples of mild steel might vary from, say, 360 to 450 MPa, whereas the quoted typical value is 400 MPa. If a particular design was based on the quoted value and it was constructed of mild steel with an actual lower value, the component would be weaker than predicted. It is against such unknown circumstances that a factor of safety is used in the analysis.

There are several definitions of the factor of safety, but that most widely used is

$$n = \frac{\sigma_u}{\sigma} \tag{8.4}$$

where $n$ = factor of safety.

$\sigma$ = actual operating stress.

The value of $n$ used depends on the actual situation. Recommended values are as follows: $n = 2$ for ductile materials where there is confidence in the material properties; $n = 3$ for brittle materials where there is confidence in the material properties; $n = 4$ for situations where there is uncertainty regarding both the loading conditions and material properties.

## Example 8.4

A cast iron cylinder of 125 mm outside diameter and 75 mm inside diameter is subjected to an axial compressive force (Figure 8.9.) Determine the maximum permitted force if the cylinder is to operate with a factor of safety of 3 and the ultimate stress in compression for cast iron is 240 MPa.

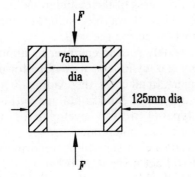

**Fig. 8.9**

*Solution*

The permitted stress can be found from equation (8.4):

$$n = \frac{\sigma_u}{\sigma}$$

and

$$\sigma = \frac{\sigma_u}{n} = \frac{240 \times 10^6}{3} = 80 \times 10^6 \, \text{Pa}$$

The permitted load can be found from equation (8.1):

$$\sigma = \frac{F}{A}$$

and

$$F = \sigma \times A$$

but

$$A = \frac{\pi}{4}(d_o^2 - d_i^2)$$

$$= \frac{\pi}{4} \frac{(125)^2 - (75)^2}{10^6} = 7.85 \times 10^{-3} \, \text{m}^2$$

Therefore

$$F = 80 \times 10^6 \times 7.85 \times 10^{-3}$$

$$= 628\,318.5 \, \text{N} = 628.3 \, \text{kN}$$

## 8.5 THIN CYLINDERS

The foregoing discussion of direct stress and strain, in which loading is in the axial direction along a component, has been considered as applying to straight bars. However, there are other types of component that are subject to direct loading and a thin cylinder is one of them.

A typical thin cylinder is shown in Figure 8.10. It is defined as a thin cylinder because the thickness of the cylinder wall is small in relation to the cylinder diameter. This allows the assumption that the stresses in the cylinder wall are constant across the wall.

Thin cylinders are widely used in engineering applications and typical examples range from gas cylinders, used for storing compressed gas, through to steam boilers. In all cases, the loading on a thin cylinder is due to the internal pressure of the fluid contained within the cylinder. This pressure causes two types of stress to develop in the cylinder wall, namely 'longitudinal' stress and 'circumferential' stress.

The pressure, $P$, inside a cylinder will be uniform over the whole of the internal surface and will act normal, in other words – at right angles to the surface. The effect of pressure on the ends of a thin cylinder causes a longitudinal stress in the cylinder wall, as illustrated in Figure 8.11. Taking a section across a cylinder, the section will be in equilibrium, so that the resultant force, due to the pressure on the end of the cylinder, must be balanced by an equal, and opposite, force in the wall:

$$P \times \frac{\pi}{4} d^2 = \sigma_L \times \pi dt$$

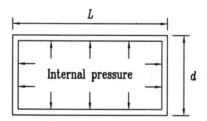

**Fig. 8.10** Loading of a thin cylinder.

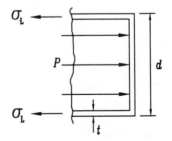

**Fig. 8.11** Longitudinal stress in a thin cylinder.

where $\sigma_L$ is the longitudinal stress in the cylinder wall. Rearranging the above relationship gives the equation for the longitudinal stress as

$$\sigma_L = \frac{Pd}{4t} \qquad (8.5)$$

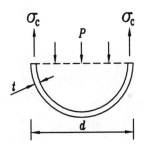

**Fig. 8.12** Circumferential stress in a thin cylinder.

Equation (8.5) indicates that the longitudinal stress increases directly with the internal pressure and the cylinder diameter, but reduces with an increase in wall thickness.

The effect of pressure on the main body of a thin cylinder causes a circumferential stress, sometimes referred to as a 'hoop' stress, in the cylinder wall, as illustrated in Figure 8.12. Taking an axial section along the centre line of the cylinder, the section will be in equilibrium, so that the resultant force due to the pressure on the projected cross-section of the cylinder must be balanced by equal, and opposite, forces in the two walls:

$$P \times d \times L = \sigma_C \times L \times 2t$$

where $\sigma_C$ is the circumferential stress in the cylinder walls. Rearranging the above relationship gives the equation for the circumferential stress as

$$\sigma_C = \frac{Pd}{2t} \qquad (8.6)$$

Comparing equations (8.5) and (8.6) it will be seen that the circumferential stress is twice the value of the longitudinal stress. This raises the interesting question of what stress applies in the case of a thin sphere subjected to an internal stress. Although the solution is given in Example 8.6 below, it should be fairly straightforward to work out the answer before turning to the solution.

**Example 8.5**

In the cylinder of a foot pump, air is compressed to a pressure of 500 kPa. If the cylinder has a diameter of 55 mm and a wall thickness of 2.5 mm, determine the stresses in the cylinder wall.

*Solution*

The longitudinal stress can be found using equation (8.5):

$$\sigma_L = \frac{Pd}{4t} = \frac{500 \times 10^3 \times 0.055}{4 \times 0.0025}$$

$$= 2\,750\,000\,\text{Pa} = 2.75\,\text{MPa}$$

The circumferential stress can be found using equation (8.6):

$$\sigma_C = \frac{Pd}{2t} = \frac{500 \times 10^3 \times 0.055}{2 \times 0.0025}$$

$$= 5\,500\,000\,\text{Pa} = 5.5\,\text{MPa}$$

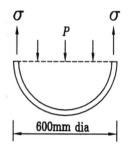

**Fig. 8.13**

**Example 8.6**

A spherical container has a diameter of 600 mm and contains nitrogen at a pressure of 2 MPa (Figure 8.13). If the allowable stress in the wall is 100 MPa, find the wall thickness.

*Solution*

Applying equilibrium to the section shown in Figure 8.13:

upward forces = downward forces

$$\sigma \times \pi \times d \times t = P \times \frac{\pi}{4} d^2$$

and

$$\sigma = \frac{Pd}{4t}$$

Rearranging

$$t = \frac{Pd}{4\sigma} = \frac{2 \times 10^6 \times 0.6}{4 \times 100 \times 10^6}$$

$$= 3 \times 10^{-3} \, \text{m} = 3 \, \text{mm}$$

*Note:* the longitudinal and circumferential stresses are the same in a sphere.

## 8.6 LATERAL STRAIN

When a bar is subjected to direct loading there is a corresponding change in the length of the bar. As discussed in section 8.3.3, the axial extension is governed by Hooke's law providing that the material is within the elastic region. The axial extension is also accompanied by a corresponding change in the lateral dimensions. This can be visualized by stretching a rubber band. The band increases in length but undergoes a noticeable reduction in thickness, as shown in Figure 8.14.

The reduction in thickness is determined by the **lateral strain**, the reduction in thickness compared to the original thickness. Now the lateral strain, $\varepsilon_L$, is proportional to the longitudinal strain, $\varepsilon$. However, the lateral strain is in the opposite sense to the longitudinal strain. If a bar is being extended in an axial direction, then it would be reduced in the lateral direction; $\varepsilon_L$ is negative when $\varepsilon$ is positive. Similarly, if a bar is being

**Fig. 8.14** Lateral contraction due to axial extension.

compressed, it would be increased in the lateral direction; $\varepsilon_L$ is positive when $\varepsilon$ is negative. This can be summarized in the relationship:

$$\varepsilon_L = -v \times \varepsilon \qquad (8.7)$$

where $v$ is the ratio of the lateral to the longitudinal strain, termed 'Poisson's ratio'.

Although equation (8.7) shows Poisson's ratio to be a negative quantity, it is always quoted as a positive value. If the material is assumed to be homogeneous, in other words, if it has perfectly uniform composition throughout, the elastic properties can be considered as the same in all directions. Therefore, the lateral strain will be the same in all directions. On this basis, the value of Poisson's ratio would be 0.5 to ensure there is no change in volume. This is demonstrated in Example 8.7 below. In reality, most engineering materials have Poisson ratio values in the range 0.25–0.35. Rubber has a value approaching 0.5, while cork has a value approaching zero. This is why cork is an ideal material for sealing wine in bottles: the diameter changes little as the cork is forced into the neck of the bottle.

## Example 8.7

A circular bar, having a length of 100 mm and diameter of 20 mm, is subjected to an axial strain of $0.4 \times 10^{-3}$. If Poisson's ratio is 0.5, show that the change of volume is zero.

*Solution*

The change in length can be found using equation (8.2):

$$\varepsilon = \frac{\Delta L}{L}$$

and

$$\Delta L = \varepsilon \times L = 0.4 \times 10^{-3} \times 100 = 0.04 \, \text{mm}$$

The new length is 100.04 mm.
    The lateral strain can be found using equation (8.7):

$$\varepsilon_L = -v \times \varepsilon = -0.5 \times 0.4 \times 10^{-3}$$
$$= -0.2 \times 10^{-3}$$

The change in diameter is
$$\Delta d = \varepsilon_L \times d = -0.2 \times 10^{-3} \times 20$$
$$= 0.004 \, \text{mm}$$

The new diameter is 19.996 mm.
    The original volume of the bar is

$$100 \times \frac{\pi}{4} \times (20)^2 \times 31\,415.9 \, \text{mm}^3$$

The new volume of the bar is

$$100.04 \times \frac{\pi}{4} \times (19.996)^2 = 31\,415.9\,\text{mm}^3$$

i.e. there is no change of volume.

### Example 8.8

A rectangular bar 40 mm wide, 10 mm thick and 500 mm long is subjected to a tensile force of 50 kN. If the bar is made of aluminium with a modulus of elasticity $E = 70\,\text{GPa}$ and Poisson's ratio $v = 0.33$, determine the changes in the linear dimensions of the bar.

*Solution*

The tensile stress can be found using equation (8.1):

$$\sigma = \frac{F}{A}$$

where the cross-sectional area is

$$\frac{40 \times 10}{10^6} = 4 \times 10^{-4}\,\text{m}^2$$

Therefore

$$\sigma = \frac{50 \times 10^3}{4 \times 10^{-4}} = 125 \times 10^6\,\text{Pa}$$

$$= 125\,\text{MPa}$$

The longitudinal strain can be found using equation (8.3):

$$\varepsilon = \frac{\sigma}{E} = \frac{125 \times 10^6}{70 \times 10^9} = 1.786 \times 10^{-3}$$

The lateral strain can be found using equation (8.7):

$$\varepsilon_L = -v \times \varepsilon = -0.33 \times 1.786 \times 10^{-3}$$

$$= -5.89 \times 10^{-4}$$

The changes in linear dimensions are:

$$\text{change in length} = \varepsilon \times \text{original length}$$
$$= 1.786 \times 10^{-3} \times 500 = 0.893\,\text{mm}$$
$$\text{change in width} = \varepsilon_L \times \text{original width}$$
$$= -5.89 \times 10^{-4} \times 40 = -0.0236\,\text{mm}$$
$$\text{change in thickness} = \varepsilon_L \times \text{original thickness}$$
$$= -5.89 \times 10^{-4} \times 10 = -0.00589\,\text{mm}$$

*Note:* the Poisson's ratio of 0.33 results in the bar increasing in volume due to this loading.

## 8.7 SHEAR LOADING

The type of loading considered in the preceding sections concerns direct loading, in which the applied force is parallel to the longitudinal axis of the bar or component being analysed. The direct loading results in direct stress and strain in the longitudinal direction, with a corresponding lateral strain.

This can be visualized by considering a piece of paper being pulled at both ends. If the applied force, the pull, is great enough, the paper will tear. There is, however, another way to tear the paper – by ripping across the paper normal to its longitudinal axis. If the longitudinal pull results in a direct stress in the paper, the ripping action across the paper results in a **shear** stress in the paper.

In Chapter 9, on the loading of beams, one form of loading is due to the 'shear' forces acting on a beam. The shear forces create shear stresses within the beam. This represents one practical example of the action of shear stresses. Another example is illustrated in Figure 8.15.

Figure 8.15(a) shows a pinned joint, subject to axial loading through an applied force, $F$. Such a joint is used widely in engineering and typical examples can be seen in mechanical braking systems and trailer hitches. Similar types of joint, operating on the same basic principle, use rivets or bolts to carry the load from one side of the joint to the other.

Under the action of the applied force the loading on the pin will be as shown in Figure 8.15(b), resulting in the pin being subjected to a shearing action across sections AB and CD. The shear forces acting on sections AB and CD will be equal to $\frac{1}{2}F$ since the total force is being carried on the two sections in parallel.

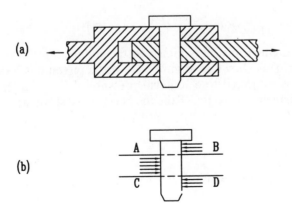

**Fig. 8.15** Shear in a bolted joint.

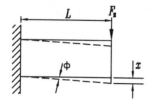

**Fig. 8.16** Shear force acting on a bar.

The type of loading shown in Figure 8.15(b) is an example of 'direct shear', sometimes referred to as 'simple shear', in which the shear stresses are created as a result of the forces trying to cut through the material. Direct shear applies to rivets and bolts, as mentioned above, and also to welded and glued joints. In order to analyse such situations, it is necessary to define shear stress and the accompanying shear strain.

### 8.7.1 Shear stress and strain

When a bar, or component, is subjected to direct loading, there will be small changes in the dimensions defined by the longitudinal and lateral strains. However, the general shape of the bar, or component, remains the same. A straight circular bar still remains a straight circular bar under direct loading. However, this is no longer true for components subjected to a shear force. A shear force causes the shape to change, as illustrated in Figure 8.16.

Figure 8.16 shows a bar fixed at one end and subject to a shear force. The shear stress in the bar can be found by dividing the shear force, $F_s$, by the area, $A$, over which it acts:

$$\tau = \frac{F_s}{A} \tag{8.8}$$

Due to the action of the shear force the bar will be deflected until it reaches a new stable position in which the free end of the bar has moved through a distance of $x$. The deflection is clearly dependent on the magnitude of the shear stress and the length of the bar, $L$. By relating the deflection to the length, it is possible to express the distortion in terms of an angle:

$$\phi = \frac{x}{L} \tag{8.9}$$

where the angle $\phi$ defines the **shear strain**, measured in radians.

### Example 8.9

In a petrol engine a piston is joined to the connecting rod by means of a gudgeon pin, as shown in Figure 8.17. If the piston has a diameter of 60 mm and is subjected to a uniform pressure of 1.5 MPa, estimate the minimum diameter of the pin. Take the permissible shear stress of the pin to be 50 MPa.

*Solution*

The total load on the piston is $P \times$ area of piston

$$= 1.5 \times 10^6 \times \frac{\pi}{4} \times \frac{60^2}{10^6} = 4241.2 \, \text{N}$$

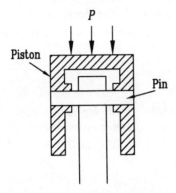

**Fig. 8.17**

The load is carried by the gudgeon pin over two sections in parallel, so that the load over one section is 2120.6 N.

The size of the pin can be found using equation (8.8):

$$\tau = \frac{F_s}{A}$$

or

$$A = \frac{\pi}{4}d^2 = \frac{F_s}{\tau}$$

$$= \frac{2120.6}{50 \times 10^6} = 4.24 \times 10^{-5}\,\mathrm{m}^2$$

The diameter of the pin is

$$d = \sqrt{\left(\frac{4 \times 4.24 \times 10^{-5}}{\pi}\right)} = 7.35 \times 10^{-3}\,\mathrm{m}$$

$$= 7.35\,\mathrm{mm}$$

### 8.7.2 Shear modulus

In direct loading the stress and strain can be related, within the elastic region, by means of the modulus of elasticity, $E$. A similar relationship can be used to relate the shear stress and shear strain, since the two are proportional to one another within the elastic region. The equivalent modulus of elasticity in shear is defined by

$$G = \frac{\tau}{\phi} \tag{8.10}$$

where $G$ is termed the **shear modulus** with units of Pa.

As with the modulus of elasticity, $G$ is a measure of the stiffness of the material and typical values are given in Appendix A for various metals.

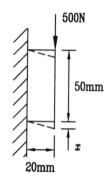

500N

50mm

20mm

**Fig. 8.18**

The modulus of elasticity, $E$, and the shear modulus, $G$, are both elastic properties of a particular material. It can be shown that they are related through the equation

$$G = \frac{E}{2(1 + v)}$$

where $v$ is Poisson's ratio. Since Poisson's ratio for most engineering materials lies between one-quarter and one-half, it follows that the value of $G$ must be between 33 and 40% of the value of $E$.

**Example 8.10**

An engine support can be modelled as a rubber pad, 50 mm square by 20 mm thick, as shown in Figure 8.18. If the support is subjected to a shear force of 500 N and $G$ for the material is 1 MPa, calculate the deflection.

*Solution*

The shear stress can be found using equation (8.8):

$$\tau = \frac{F_s}{A} = \frac{500 \times 10^6}{50 \times 50} = 200\,000 \,\text{Pa}$$

$$= 200 \,\text{kPa}$$

The shear strain can be found using equation (8.10):

$$G = \frac{\tau}{\phi}$$

and

$$\phi = \frac{\tau}{G} = \frac{200 \times 10^3}{10^6} = 0.2$$

The deflection can be found using eqution (8.9):

$$\phi = \frac{x}{L}$$

and

$$x = \phi \times L = 0.2 \times 20 = 4 \,\text{mm}$$

**8.8 SUMMARY**

Key equations that have been introduced in this chapter are as follows.
    Direct stress:

$$\sigma = \frac{F}{A}$$

(8.1)

Direct strain:

$$\varepsilon = \frac{\Delta L}{L} \tag{8.2}$$

Elastic relationship for direct loading:

$$E = \frac{\sigma}{\varepsilon} \tag{8.3}$$

Factor of safety:

$$n = \frac{\sigma_u}{\sigma} \tag{8.4}$$

Stresses in thin cylinders – longitudinal stress:

$$\sigma_L = \frac{Pd}{4t} \tag{8.5}$$

and circumferential, or hoop, stress:

$$\sigma_C = \frac{Pd}{2t} \tag{8.6}$$

Lateral strain under the action of direct loading:

$$\varepsilon_L = -v \times \varepsilon \tag{8.7}$$

Shear stress:

$$\tau = \frac{F_s}{A} \tag{8.8}$$

Shear strain:

$$\phi = \frac{x}{L} \tag{8.9}$$

Elastic relationship for shear loading:

$$G = \frac{\tau}{\phi} \tag{8.10}$$

## 8.9 PROBLEMS

1. A structural member has a rectangular cross-section, 80 by 30 mm, and carries a tensile load of 100 kN. What is the stress in the member? If it is made of steel with a modulus of elasticity of 200 GPa, what is the strain?
2. A tie bar 2 m in length has a circular cross-section of 22 mm diameter and carries a longitudinal load of 40 kN. Calculate the stress in the bar and the change in length if $E = 200$ GPa.
3. A stepped circular bar is subjected to an axial tensile force of 500 kN. If the bar consists of a 100 mm diameter section, 800 mm long, and

a 60 mm diameter section, 1 m long, find:

(a) the stress in each section;

(b) the change in length of the bar if $E = 200$ GPa.

4. An aluminium bicycle pump has a diameter of 25 mm and a wall thickness of 1 mm. Determine the circumferential stress in the wall if the internal pressure is 500 kPa.

5. A cylindrical tank, to be used with an air compressor, is 1.2 m long and 500 mm in diameter. If the wall of the tank is 3 mm thick and the stress in the wall is not to exceed 100 MPa, find the maximum permissible pressure in the tank.

6. A brass rod 25 mm in diameter and 600 mm long carries an axial load of 42 kN. Measurement shows that the diameter reduces by 0.0067 mm and the length extends by 0.63 mm. Determine the modulus of elasticity and Poisson's ratio for the material.

7. A rectangular bar 50 mm wide, 12 mm thick and 800 mm long is subjected to a tensile force of 75 kN. If the bar is made of steel with $E = 200$ GPa and $v = 0.3$, calculate the changes in the dimensions.

8. A loose-fitting bar fits inside a tube. The two are joined by means of a 6 mm diameter bolt normal to the longitudinal axis. If the axial force is 1.6 kN, find the shear stress in the bolt.

9. A draw bar between a tractor and trailer is made from a length of steel bar of 80 mm by 12 mm rectangular cross-section. The load is transmitted by means of a 15 mm diameter pin at each end. Determine the maximum stresses in the bar and pin if the axial load is 10 kN.

10. Circular holes are punched out of 5 mm thick steel plate using a 40 mm diameter punch. If the ultimate shear stress of the steel plate is 450 MPa, determine the compressive stress in the punch.

# Loading of beams $\boxed{9}$

## 9.1  AIMS

- To introduce and define beam types and their support conditions.
- To describe the types of load which may be applied to horizontal beams.
- To explain how any arbitrary position on a beam must be subject to shear force and bending moment in order to achieve equilibrium.
- To describe the construction of shear force and bending moment diagrams for a beam.
- To introduce a method of analysis for complex loading of beams.

## 9.2  SIMPLY SUPPORTED BEAMS AND CANTILEVERS

Beams are important structural members which usually support transverse loads. Such members are found in bridges, roof trusses and cranes. Normally the beam is horizontal and the forces vertical, though this may not always be the case.

Forces which act on a beam may be:

1. Loads acting downwards as a result of masses supported by the beam and which can be considered as concentrated loads since each load will act approximately at a single point.
2. Loads which are spread over part or the full length of the beam – called **uniformly distributed loads** or (UDLs). Examples of UDLs are: the self-weight of a beam, a lintel over the window of a house, holding up the brickwork or a beam supporting a long water tank.
3. The reactions which support the beam and all its loads.

Beams are often classified by their idealized support conditions (reactions), two of which are shown in Figures 9.1 and 9.2.

A **simply supported beam** always has two supports, often one at each end, on which the beam rests. It is not rigidly constrained and can bend (deflect) freely if loads are applied.

A **cantilever** is a beam which is supported at one end only and must be 'built in', perhaps by building a short length into masonry or concrete or

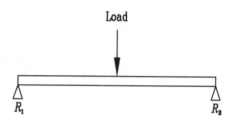

**Fig. 9.1** Simply supported beam.

**Fig. 9.2** Cantilever.

by welding it to a rigid structure. There is a dual reaction at the built-in end, consisting of:

1. an upwards vertical reaction which supports the weight of the beam and any downwards loads, plus,
2. a moment, called a **fixing moment**, which prevents the unsupported length of the beam from falling.

### 9.2.1 Equilibrium of a loaded beam

When a beam is loaded it will tend to bend. The analysis of this behaviour requires a knowledge of how the internal forces and moments respond to the applied loads and support reactions. If the beam is stationary (static) under all loads and reactions, then a state of equilibrium must exist and this can be used as a basis for analysis.

The conditions for equilibrium apply not only to the whole beam but to every part of it.

Consider a beam under load, which is cut in half and then welded together. The welded joint will have to cope with a number of forces and moments in order to hold the beam together in equilibrium.

Figure 9.3(a) shows a cantilever under two loads $W_1$ and $W_2$ which is sectioned at Z–Z across the horizontal plane. Figure 9.3(b) shows how the forces and moments act on each side of the plane.

In Figure 9.3(b) it can be observed that if part B is to remain in equilibrium the force $F$ must counteract the load $W_2$. The force $F$, however, is not enough to prevent this portion of the beam from rotating. Moment $M$ must therefore be present to ensure equilibrium.

It should be noted that it is part A of the beam which supports part B. This means that on the left-hand side of the cutting plane there must be

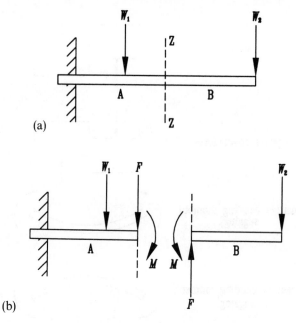

**Fig. 9.3** Cutting plane on a beam showing forces and moments which ensure equilibrium.

equal and opposite forces and moments, $F$ and $M$, which compensate for those forces and moments on the right-hand side of the cutting plane.

Thus it can be said that at the junction of the Z–Z plane the two parts of the cantilever exert on each other equal and opposite forces $F$ and equal and opposite moments $M$. These are normally transmitted by forces within the beam, which result in stresses within the material.

The forces $F$ are known as **shear forces** since they tend to 'cut' or 'shear' the beam and are responsible for 'shear stresses'. The moments $M$ are known as **bending moments** and give rise to 'bending stresses'.

In designing a beam it is important to know the position of the highest stresses. These correspond to the values of the highest shear stress/bending moment combinations.

The shear force (SF) and bending moments (BM) vary along the length of the beam, and this variation can be shown by drawing graphs in which the beam is represented by the horizontal axis and the values of SF and BM are plotted vertically. These diagrams are called shear force and bending moment diagrams respectively.

## 9.3 SIGN CONVENTIONS

Before discussing the SF and BM diagrams in detail it is necessary to establish a sign convention.

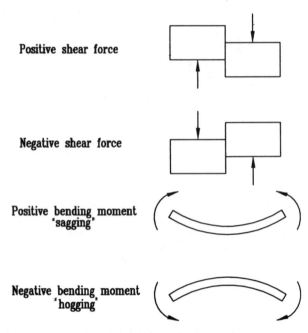

Positive shear force

Negative shear force

Positive bending moment 'sagging'

Negative bending moment 'hogging'

**Fig. 9.4** Sign conventions of beams under the influence of shear forces and bending moments.

Since SFs may be applied upwards or downwards on a beam and moments may try to turn the beam either clockwise or anticlockwise, it is necessary to select a convenient system where certain directions of forces and moments may be taken as positive. It is an arbitrary choice as to which directions are taken as positive, but it is important to be consistent throughout any analysis.

It is suggested that the practice outlined below is used since it follows general mathematical conventions.

Figure 9.4 shows beams under the influence of various SFs and BMs and illustrates the sign conventions used in this text:

- positive shear force: down and to the right of the section;
- positive bending moment: anticlockwise moment to the right of the section.

Alternatively: the positive moments cause the beam to sag.

## 9.4 SHEAR FORCES AND SHEAR FORCE DIAGRAMS

### 9.4.1 Shear force diagrams for concentrated loads

The shear force at any point along a beam is the algebraic sum of all the normal forces acting on either side of the beam.

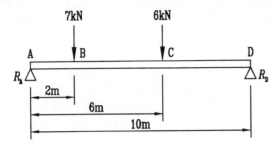

**Fig. 9.5** Space diagram.

This definition may be applied to any beam and in practice means that the total SF may be found at any point by progressing along the beam from one end and adding positive forces and subtracting negative forces. This is illustrated in Example 9.1.

**Example 9.1**

Figure 9.5 shows a beam ABCD, which is simply supported at each end and is acted upon by two concentrated loads. Construct the SF diagram for the beam.

*Solution*

Before the SF diagram can be found it is first necessary to determine the reactions at the supports A and D.

**Find reaction $R_D$**

Taking moments about A:

$$\circlearrowleft = \circlearrowright$$

$$R_D \times 10 = (7 \times 2) + (6 \times 6)$$

$$R_D = \frac{14 + 36}{10}$$

$$= 5.0 \, \text{kN}$$

**Find reaction $R_A$**

For equilibrium of vertical forces:

$$R_A + 5 = 7 + 6$$

$$R_A = 8.0 \, \text{kN}$$

**Find shear forces**

The SF diagram may be built up as follows:

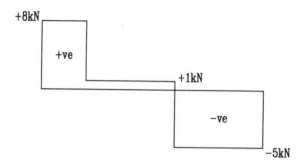

**Fig. 9.6** Shear force diagram.

1. Starting from end A and progressing to the right select a point on the beam usually at the point of application of the next force, in this case point B. The reaction $R_A$ is 8.0 kN and is the positive SF which acts on the beam as far as point B. This may now be drawn on the SF diagram as a positive 8.0 kN.
2. At point B a negative force of 7.0 kN is algebraically added to the SF diagram and results in a total positive SF of 1.0 kN. This acts on the beam as far as the next force application at point C.
3. At point C another negative force of 6.0 kN is encountered and is algebraically added to the SF diagram resulting in a $-5.0$ kN SF at point C. This acts on the beam as far as point D.
4. Finally at point D the positive reaction force of $R_D = 5.0$ kN is encountered and when algebraically added to the SF diagram gives a value of zero at the end of the beam. The resulting SF diagram is shown in Figure 9.6.

The result of zero at the second support is most important because it demonstrates that:

1. All the downwards forces equal the upwards forces.
2. The beam is in vertical equilibrium.

The result of zero should always occur at the end of the beam and is an excellent check for calculations.

**Example 9.2**

Figure 9.7 shows a cantilevered beam ABC, which is acted upon by two concentrated loads. Construct the SF diagram for the beam.

*Solution*

The SF diagram can be built up as follows:

1. Start the analysis at the load which is furthest away from the built-in end. This is at point A. On the SF diagram mark off a scale length of 6 kN, vertically. Since the SF is positive the SF diagram will be drawn above the base line.

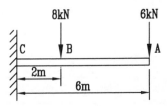

**Fig. 9.7** Cantilever beam with point loads.

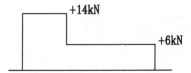

**Fig. 9.8** Shear force diagram for a cantilever with point loads.

2. Progress along the beam to the left. The 6 kN shear force is applied until point B is reached, which is the point of application of the 8 kN force. The SF at this point is

$$SF = 8 + 6 = 14\,kN$$

and can be marked on the SF diagram.
3. Progress along the beam to point C, the built-in end. The 14 kN SF is applied along the beam to the support. The SF diagram can now be completed as shown in Figure 9.8.

*Note:* The SF which closes the diagram has a value of 14 kN and acts in the opposite direction to the applied forces. If equilibrium of forces is to be present, then all the forces acting downwards must equal all the forces acting upwards. The beam is, therefore, in equilibrium since the support supplies the necessary reactive force.

### 9.4.2 Shear force diagrams for distributed loads

The concept of a point load is an idealized application of a force since in practice all loads are applied across some 'footprint'. For a point load the area of the footprint may be considered negligible. There are cases, however, where a distributed load is significant such as in the self-weight of a beam. Other examples include a lintel which supports brickwork over a door, or a bookshelf loaded with uniformly sized books.

### Example 9.3

Figure 9.9 shows a simply supported beam acted upon by a uniformly distributed load (UDL) of 6.0 kN/m.

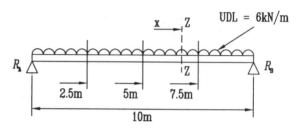

**Fig. 9.9** Simply supported beam with a uniformly distributed load.

(a) Determine the reactions at the supports.
(b) Construct the SF diagram for the beam.

*Solution*

(a) The reactions at the supports can be found as follows:

total load $= w \times l = 6.0 \,\text{kN/m} \times 10.0 \,\text{m} = 60.0 \,\text{kN}$

where

$w = \text{UDL} = 6.0 \,\text{kN/m}$
$l = \text{total length} = 10 \,\text{m}$

Since the load is distributed evenly between the supports $R_A$ and $R_B$, each support will have a value of half the total load, i.e. 30.0 kN.

(b) For the UDL shown in Figure 9.9, the SF can be calculated as follows:

   (i) Start at the reaction $R_A$ which has a value of 30.0 kN, and mark this on the SF diagram (Figure 9.10).

   (ii) Progress along the beam to the right and take a section at, say, 2.5 m from $R_A$. The UDL along this length is acting down, i.e. is negative and has a value of

   $2.5 \times 6.0 = 15.0 \,\text{kN}$

   Algebraic addition of positive and negative forces gives

   $30.0 - 15.0 = 15.0 \,\text{kN}$

   at this point.

   (iii) Progress along the beam to the right and take another section at, say, 5.0 m from $R_A$. Perform the same calculation as before. The UDL is acting in a negative direction and has a value of

   $5.0 \times 6.0 = 30.0 \,\text{kN}$

   The algebraic addition of positive and negative forces gives

   $30.0 - 30.0 = 0 \,\text{kN}$

   at this point.

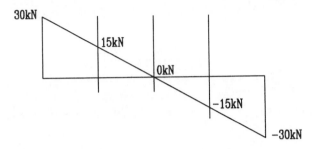

**Fig. 9.10** Shear force diagram for a simply supported beam acted upon by a uniformly distributed load.

(iv) Progressing along the beam to, say, 7.5 m from the left-hand end, take another section. The UDL is acting in a negative direction and has a value of

$$7.5 \times 6.0 = 45.0 \, kN$$

The algebraic addition of positive and negative forces gives

$$30.0 - 45.0 = -15.0 \, kN$$

at this point.

(v) The final section is taken at the reaction $R_B$ where the whole UDL is acting in a negative direction and has a value of

$$10.0 \times 6.0 = 60 \, kN$$

The algebraic addition of positive and negative forces gives

$$30.0 - 60.0 = -30.0 \, kN$$

The results obtained in (i)–(v) above can now be plotted on the SF diagram shown in Figure 9.10. It should be noted that:

- The SF varies along the length of the beam at a rate of 6.0 kN/m.
- The reaction at $R_B$ is 30 kN and reduces the value of the SF to zero, which results in a state of equilibrium.

Uniformly distributed loads can also be applied to cantilevered beams. The problem is considered in a similar way to that shown for point loads in Example 9.2; however, sections (hinges) are placed at convenient points along the beam, as described in Example 9.4.

### Example 9.4

Figure 9.11 shows a cantilever under the influence of a 6.0 kN/m uniformly distributed load. Construct the SF diagram for the beam.

*Solution*

The SF diagram can be constructed as follows:

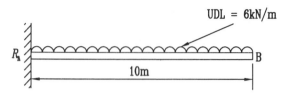

**Fig. 9.11** Cantilevered beam acted upon by a uniformly distributed load.

1. Start at the unsupported end where the load is zero.
2. Move towards the built-in end an arbitrary but convenient length, say 2.0 m. The SF at this point is that load supported by the beam to the right, i.e.

load $= 6 \times 2.0 = 12\,\text{kN}$

3. Move towards the built-in end another 2.0 m, which is 4.0 m in total from the unsupported end. The SF at this point is that load supported by the beam to the right, i.e.

load $= 6 \times 4.0 = 24\,\text{kN}$

4. Take further points, at 2.0 m intervals, towards the built-in end, and at each interval perform the same operation as that shown in (2) and (3) above. The SF values are developed as follows.
   At 6.0 m from the unsupported end:

load $= 6 \times 6.0 = 36\,\text{kN}$

At 8.0 m from the unsupported end:

load $= 6 \times 8.0 = 48\,\text{kN}$

At 10.0 m from the unsupported end:

load $= 6 \times 10.0 = 60\,\text{kN}$

Note that the maximum shear force is 10.0 m from the unsupported end. This, of course, is the position of the built-in support.

The SF diagram can now be drawn as shown in Figure 9.12.

The cantilever can be analysed using a number of observations which can provide a short-cut for the analysis, as follows:

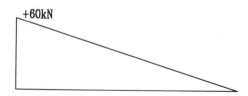

**Fig. 9.12** Shear force diagram.

- The SF diagram must be zero at point B.
- There is no reaction at point B.
- The UDL applies a uniform slope of 6.0 kN/m to the graph.
- The maximum SF value is always at the built-in end.

The conclusion is that the value of the SF diagram gradually reduces from 60.0 kN to zero as progression is made along the beam from $R_A$ to B. In this case the SF diagram never crosses the base line.

## 9.5 BENDING MOMENTS AND BENDING MOMENT DIAGRAMS

### 9.5.1 Bending moments and bending moment diagrams for concentrated loads

Any beam acted upon by a number of point loads will tend to bend and will, therefore, be subjected to bending moments (BMs).

A BM is the result of a force being applied at a distance from an assumed pivot point or hinge, and is a numerical value which describes the 'intensity of bend' at that point. The analysis presented below shows how BMs can be calculated along the entire length of a beam, and is the preliminary analysis required before the stress can be found at any point along the length of the beam.

The beam shown in Figure 9.13 is the same as that used in Example 9.1 where the reactions at the supports were calculated prior to the SFs.

The beam has a clockwise BM imposed on it by the 8.0 kN reaction. The Z–Z plane has been arbitrarily placed 1.0 m from the left-hand end. If the beam were to be sliced in two at this point, the 8.0 kN reaction would apply a clockwise BM of

$$8.0\,kN \times 1.0\,m = 8.0\,kN\,m$$

The position of the Z–Z plane is, therefore, important because it becomes an imaginary 'hinge' around which the beam may wish to bend.

During the calculation of BMs it is necessary to progress along the beam from the left-hand end, in a similar manner to that employed for SF

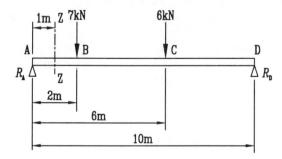

**Fig. 9.13** Simply supported beam acted upon by concentrated loads.

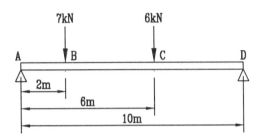

**Fig. 9.14** Simply supported beam acted upon by concentrated loads.

calculations. It is usual to place the hinge at each point load, such as at point B, where the 7.0 kN load is applied. This eliminates that load since the absence of a moment arm eliminates any moment caused by that load.

**Example 9.5**

Construct the BM diagram for the simply supported beam shown below in Figure 9.14. This is the same beam and loading system as used in Example 9.1 and the reactions have previously been calculated as $R_A = 8\,kN$ and $R_D = 5\,kN$.

*Solution*

The BM diagram can be calculated and constructed as follows:

1. Starting from the left-hand end progress along the beam to point B, where the 7.0 kN load is applied. To illustrate this situation it is helpful to place the point of a pen on the diagram.
   This is the hinge point and only forces to the left of this point can be considered.
2. The only force creating a moment is the reaction of 8.0 kN. The moment is positive and is calculated as follows:

   $M_B =$ force $\times$ distance
   $\quad\ \ = 8 \times 2 = 16.0\,kN\,m$

   This can be marked on the BM diagram.
3. Progress along the beam to point C, where the 6.0 kN load is applied. This is the second hinge point and only forces to the left can be considered.
4. There is a positive moment applied at A, caused by the 8.0 kN load, and a negative moment applied at B, caused by the 7.0 kN load. Sum the moments as follows:

   $M_C = (8 \times 6) - (7 \times 4) = 20.0\,kN\,m$

   This can now be applied to the BM diagram.
5. Where the reactions are placed at each end of the beam, the BM at the second reaction should always be zero. This shows that the beam is in equilibrium.

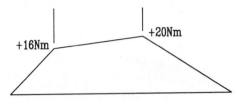

**Fig. 9.15** Bending moment diagram.

The BM diagram shown in Figure 9.15 can be drawn using the values derived above.

6. It is good practice to check calculations by placing the hinge at point D, where the 5.0 kN reaction is applied, and considering all the forces to the left of the hinge. If the calculations have been performed correctly, the BM at D should be zero.

7. The BM is calculated as follows:

$$M_D = (8 \times 10) - (7 \times 8) - (6 \times 4) = 0\,\mathrm{kN\,m}$$

8. It may be concluded that the BM calculations are correct and since the BM diagram is positive the beam will sag.

### 9.5.2 The calculation of bending moments for cantilevered beams acted upon by concentrated loads

Cantilevered beams are often used in practice and can have various loading systems applied to them. Example 9.6 below shows how the BM diagram can be derived for a cantilever acted upon by concentrated loads.

**Example 9.6**

Determine the BM diagram for the cantilever shown in Figure 9.16.

*Solution*

The BM diagram can be constructed as follows:

1. Start at the load furthest from the built-in end. In this case it is the 6.0 kN load at point A. If the hinge point were to be placed here, there would be zero BM to the right, since there is no moment arm.

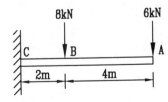

**Fig. 9.16** Cantilever acted upon by point loads.

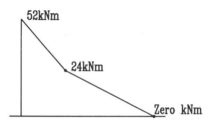

**Fig. 9.17** Bending moment diagram for a cantilever with applied point loads.

2. Move towards the built-in end and place the hinge point at point B, the point of application of the 8.0 kN load. Only loads to the right of this point can be considered. The moment arm is, therefore, 4.0 m and the BM can be calculated thus:

$$M_B = 6 \times 4 = 24 \,\text{kN m}$$

3. Move the hinge to point C, the built-in end. Only loads to the right of this point can be considered, which is in fact the whole beam. The BM here is a combination of the 8.0 kN load applied over 2.0 m and the 6.0 kN load applied over 6.0 m, and can be calculated as follows:

$$M_C = (6 \times 6) + (8 \times 2) = 52 \,\text{kN m}$$

These values can now be applied to the BM diagram (Figure 9.17). It should be noted that:

- The BM is always zero at the unsupported end.
- The BM is always a maximum at the built-in end.
- The maximum BM is always counteracted by the unseen moment within the support.

### 9.5.3 Simply supported beams under the influence of uniformly distributed loads

Beams under the influence of uniformly distributed loads (UDLs) can be treated in a similar manner to the beams under the influence of point loads; however, since there are no point loads at which a hinge may be applied, convenient places along a beam should be selected. These may be chosen arbitrarily, perhaps at 1.0 or 2.0 m intervals. The closer these intervals, the more accurate the BM diagram.

The method of analysis is described in more detail in the following example, which uses the beam originally shown in Example 9.3.

**Example 9.7**

Construct the BM diagram for the beam shown in Figure 9.18.

*Solution*

The presence of a UDL means that 'artificial' point loads need to be created in order to calculate the moments according to force × distance.

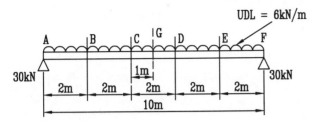

**Fig. 9.18** Simply supported beam under the influence of a uniformly distributed load.

Hinges have to be placed at convenient points and in this case these intervals have been selected at every 2.0 m. The same calculation procedure applies as before in that analysis starts at the left-hand end and progresses to the right along the beam.

The first hinge is, therefore, placed at point B which is 2.0 m from the left-hand end. The artificial point load can be taken to act at the centre of this short span and its value is

$$(6.0\,\text{kN/m}) \times (\text{the span of } 2.0\,\text{m})$$

shown in Figure 9.19.

Using this technique the analysis is performed as follows:

1. Place the hinge at point B.
2. The BM applied by the UDL alone can be calculated as follows:

$$M = \text{force} \times \text{moment arm}$$

$$= (\text{UDL value} \times \text{span}) \times \frac{(\text{span})}{2}$$

$$= (6 \times 2) \times \frac{(2)}{2}$$

$$= 12.0\,\text{kN m}$$

3. The total BM applied at point B can be calculated thus:

$$M_{\text{B}} = \text{positive moments} - \text{negative moments}$$
$$= (\text{force} \times \text{moment arm}) - (\text{force} \times \text{moment arm})$$

$$= (\text{reaction} \times \text{span}) - \left((\text{UDL} \times \text{span}) \times \frac{(\text{span})}{2}\right)$$

$$= (30 \times 2) - \left((6 \times 2) \times \frac{(2)}{2}\right)$$

$$= 60 - 12 = 48\,\text{kN m}$$

This value can now be placed on the BM diagram.
4. Move the hinge to point C, 4.0 m from the left-hand end.

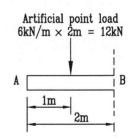

Artificial point load
6kN/m × 2m = 12kN

**Fig. 9.19** Short span of beam showing the development of the artificial point load.

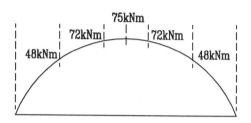

**Fig. 9.20** Bending moment diagram.

5. The BM can be calculated as follows:

$$M_C = \text{positive moments} - \text{negative moments}$$

$$= (30 \times 4) - \left( (6 \times 4) \times \frac{(4)}{2} \right)$$

$$= 120 - 48 = 72.0 \, \text{kN m}$$

This value can now be placed on the BM diagram.

6. Since the beam is symmetrical any further analysis will reveal the same BM values for the right-hand half of the beam as were calculated for the left-hand half. It should be noted, however, that a crucial hinge point has been omitted from those taken so far and this is at the centre of the span. It is of importance because it is the position of the **maximum bending moment**.

7. Take a supplementary hinge point 5.0 m from the left-hand end, at point G, and calculate the BM as follows:

$$M_G = (30 \times 5) - \left( (6 \times 5) \times \frac{(5)}{2} \right)$$

$$= 150 - 75 = 75.0 \, \text{kN m}$$

This value can be marked on the BM diagram and the whole diagram completed as shown in Figure 9.20.

8. Note the curvature of the BM line. This always occurs when a uniformly distributed load is applied.

**Example 9.8**

The cantilever shown in Figure 9.21 is acted upon by a UDL of 6 kN/m. Construct the BM diagram.

As in Example 9.7, the presence of a UDL requires the creation of an 'artificial' point load to enable analysis to proceed. Hinges have to be placed at points and in this case these intervals are every 2.0 m.

*Solution*

The BM diagram can be constructed according to the following procedure:

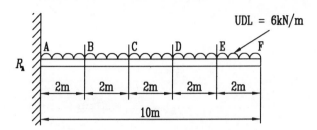

**Fig. 9.21** Cantilever subjected to a uniformly distributed load.

1. If the hinge is placed at point F, the end of the beam, there will be zero BM since there is no moment arm.
2. Progress along the beam towards the built-in end, to point E and fix the hinge. Only the beam to the right of the hinge will be considered. The 2.0 m of beam protruding to the right of the hinge carries a UDL of 6.0 kN/m. The artificial point load can be calculated as follows:

artificial point load $= 2 \times 6 = 12\,\text{kN}$

This can be taken to act from the centre of the 2.0 m span, i.e. 1.0 m from the end of the beam and 1.0 m from E. The BM at E is, therefore, calculated as follows:

$$M_E = \text{artificial load} \times \text{moment arm}$$
$$= 12 \times 1.0 = 12\,\text{kN m}$$

3. Progress along the beam towards the built-in end to point D and fix the hinge. Only the beam to the right of the hinge can be considered. The UDL is carried by the 4.0 m of beam protruding to the right of the hinge. The artificial point load is therefore calculated as follows:

artificial point load $= 4 \times 6 = 24\,\text{kN}$

This can be taken to act from the centre of the 4.0 m span, i.e. 2.0 m from D and 2.0 m from the end of the beam. The BM at D can, therefore, be calculated as follows:

$$M_D = \text{artificial load} \times \text{moment arm}$$
$$= 24 \times 2.0 = 48\,\text{kN m}$$

4. A similar approach can be taken with points A, B and C. Their respective BMs can be calculated as follows:

$$M_C = (6 \times 6) \times \frac{6}{2} = 108\,\text{kN m}$$

$$M_B = (6 \times 8) \times \frac{8}{2} = 192\,\text{kN m}$$

$$M_A = (6 \times 10) \times \frac{10}{2} = 300\,\text{kN m}$$

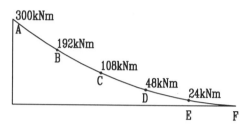

**Fig. 9.22** Bending moment diagram for a cantilever under the influence of a uniformly distributed load.

These values can now be plotted on the BM diagram as shown in Figure 9.22.

The following points should be noted:

- The maximum BM is always at the built-in end.
- There is zero BM at the unsupported end.
- When acted upon by a UDL, the BM shows a near parabolic increase towards the built-in end. This is due to the twin factors of increase in moment arm and increase in supported load as the hinge is progressed towards the built-in end.

## 9.6 THE PRINCIPLE OF SUPERPOSITION

Bending moments may be induced in a beam by any load acting alone or in combination with other loads of varying type. The determination of total BM values will lead to the calculation of the stress at any point on the beam.

The BM at any point of a structure, beam or strut carrying several loads may be found by considering each load separately as if it acted alone. The total BM is the algebraic sum of the BMs caused by each separate load.

This is called the **principle of superposition**. The loading on the beams previously shown in Examples 9.5 and 9.7 is now combined on a single beam shown in Figure 9.23(a). The point loads of 6.0 and 7.0 kN have been combined with the UDL of 6.0 kN/m. It is possible to calculate the total BM at any point on the beam, by using the method described in section 9.5.1; however, when there is a combination of point loads and UDLs the analysis soon becomes complex.

In such cases it is a simple matter to use the principle of superposition. The point loads are considered separately from the UDLs, and BM diagrams are derived for both cases, as in Examples 9.5 and 9.7. These diagrams are reproduced in Figure 9.23(b) and (c).

The total BM diagram is shown in Figure 9.23(d) and is merely the algebraic addition of the BM values from the separate diagrams in figures 9.23(b) and (c). The addition is performed at convenient points along the beam. Greater accuracy will be achieved the closer together these points are taken.

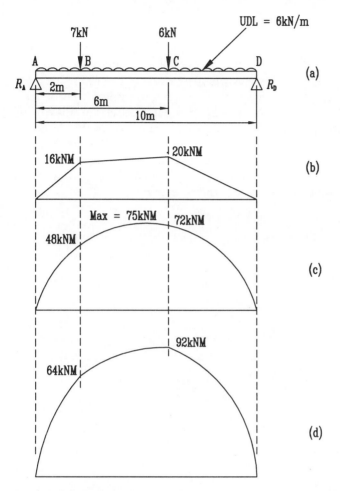

**Fig. 9.23** Simply supported beam with a combination of concentrated loads and uniformly distributed loads: (a) beam with combined loading; (b) bending moment for point loads; (c) bending moment for the uniformly distributed load; (d) combined bending moment diagram for point loads and uniformly distributed loads.

## 9.7 THE SIGNIFICANCE OF BENDING MOMENTS

It is important to emphasize the significance of BMs and in particular the maximum BM. The value of a BM along the length of a beam is proportional to the stress to which that part of the beam is subjected. In a symmetrically loaded beam the maximum BM may be judged, and can be shown, to be at the centre of the span. **Maximum stress** and **maximum deflection** also occur at this point.

The determination of the BM for any point on a beam is a first step in calculating the deflection and stress, which establishes whether a structure will fail or whether it is safe.

## 9.8 SHEAR FORCE AND BENDING MOMENT DIAGRAMS

Shear force and BM diagrams are often drawn together, since together they give an overall picture of the effects of a loading system on a beam.

In practice it is often acceptable merely to determine the maximum BM and its position, since this will enable the calculation of the maximum stress. Shear force diagrams are particularly useful in showing the position of the maximum BM, because the point at which the SF line crosses the horizontal axis is always the position of the maximum BM. The principle is shown in Figure 9.24. If a full BM picture is not required, then this method gives a quick and easy way of locating and analysing the parts of a beam which usually require the greatest scrutiny.

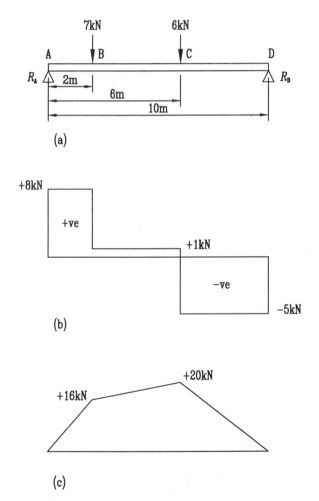

(a)

(b)

(c)

**Fig. 9.24** Combined shear force and bending moment diagrams: (a) loaded beam; (b) shear force diagram; (c) bending moment diagram.

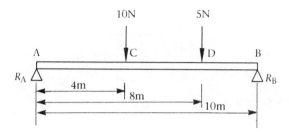

**Fig. 9.25** Simply supported beam under the influence of concentrated loads.

### Example 9.9

For the beam shown in Figure 9.25 determine:

(a) the reactions at the supports;
(b) the SF diagram;
(c) the BM diagram; and
(d) indicate the position of the maximum BM.

*Solution*

(a) The reactions at the supports can be found by initially taking moments about B, to find the reaction at A, then using the equilibrium of vertical forces to determine the reaction at B as follows.
   Moments about B:

$$\circlearrowright = \circlearrowleft$$

$$R_A \times 10 = (10 \times 6) + (5 \times 2)$$

$$R_A = \frac{70}{10} = 7\,N$$

Using equilibrium of vertical forces

$$R_A + R_B = 10 + 5$$

$$R_B = (10 + 5) - 7 = 8\,N$$

(b) The SF can be calculated as follows:

SF at A $= +7\,N$

SF at C $= 7 - 10 = -3\,N$

SF at D $= 7 - 10 - 5 = -8\,N$

SF at B $= 7 - 10 - 5 + 8 =$ zero N

   The SF diagram (Figure 9.26) can now be drawn.

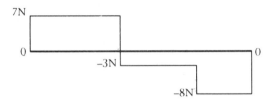

**Fig. 9.26** Shear force diagram.

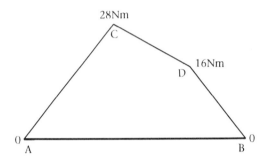

**Fig. 9.27** Bending moment diagram.

(c) The BMs can be calculated as follows:

hinge at $C = M_C = 7 \times 4 = 28 \, \text{N m}$
hinge at $D = M_D = (7 \times 8) - (10 \times 4) = 16 \, \text{N m}$
hinge at $B = M_B = (7 \times 10) - (10 \times 6) - (5 \times 2) = \text{zero N m}$

The BM diagram (Figure 9.27) can now be drawn.

(d) The maximum BM is indicated at the point where the SF diagram crosses the horizontal axis. This can be confirmed by reading the corresponding point on the BM diagram, and it can be seen that the value of 28 N m is the maximum BM for the beam.

## Example 9.10

The beam shown in Figure 9.28 has a UDL applied over part of the beam. Determine:

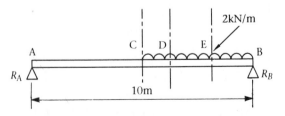

**Fig. 9.28** Simply supported beam influenced by a uniformly distributed load over part of the beam.

(a) the reactions at the supports;
(b) the SF diagram;
(c) the BM diagram.

*Solution*

(a) The reactions at the supports can be found by initially taking moments about B, to find the reaction at A, then use the equilibrium of vertical forces to determine the reaction at B as follows.
    Moments about B:

$$R_A \times 10 = (5 \times 2) \times \tfrac{5}{2}$$

$$R_A = \frac{25}{10} = 2.5 \, \text{kN}$$

Using equilibrium of vertical forces

$$R_B = (5 \times 2) - 2.5 = 7.5 \, \text{kN}$$

(b) The SF diagram can be determined using the following procedure. Fix arbitrary points C, D and E on the UDL.

SF at  A $= +2.5 \, \text{kN}$
SF at  C $=$ same as at A but is the beginning of the UDL
SF at  D $= 2.5 - (1.5 \times 2) = -0.5 \, \text{kN}$
SF at  E $= 2.5 - (3.5 \times 2) = -4.5 \, \text{kN}$
SF at  B $= 2.5 - (5 \times 2) + 7.5 = \text{zero} \, \text{kN}$

    These values can now be drawn on the SF diagram (Figure 9.29).
(c) The BM diagram can be determined by working from the left-hand end of the beam as follows:

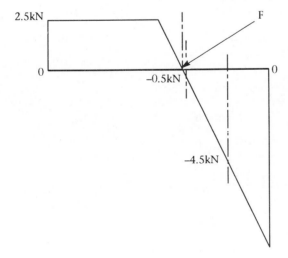

**Fig. 9.29** Shear force diagram.

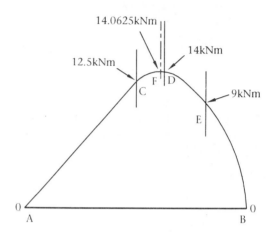

**Fig. 9.30** Bending moment diagram.

$$\text{hinge at } C = M_C = R \times 5 = 2.5 \times 5 = 12.5 \,\text{kN m}$$

$$\text{hinge at } D = M_D = (2.5 \times 6.5) - \left( (1.5 \times 2) \times \frac{1.5}{2} \right) = 14 \,\text{kN m}$$

$$\text{hinge at } E = M_E = (2.5 \times 8.5) - \left( (3.5 \times 2) \times \frac{3.5}{2} \right) = 9 \,\text{kN m}$$

$$\text{hinge at } B = M_B = (2.5 \times 10) - \left( (5 \times 2) \times \frac{5}{2} \right) = \text{zero} \,\text{kN m}$$

The maximum BM is at the point where the SF diagram crosses the horizontal axis. Fix point F at this position and calculate the maximum BM:

$$\text{hinge at } F \times M_F = (2.5 \times 6.25) - \left( (1.25 \times 2) \times \left( \frac{1.25}{2} \right) \right) = 14.0625 \,\text{kN m}$$

These values can now be drawn on the BM diagram (Figure 9.30).

### Example 9.11

The loading system on the beam shown in Figure 9.31 comprises point loads and a UDL. Determine:

(a) the reactions at the supports;
(b) the SF diagram;
(c) BM diagram; and
(d) indicate the position and magnitude of the maximum BM.

*Solution*

(a) The reactions at the supports can be found by first taking moments about A, then taking moments about D, as follows.

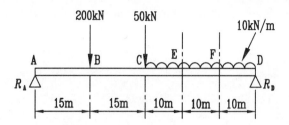

**Fig. 9.31** Simply supported beam under the influence of a combination of concentrated and uniformly distributed loads.

Moment about A:

$$\circlearrowleft = \circlearrowright$$

$$R_D \times 60 = (200 \times 15) + (50 \times 30) + ((10 \times 30) \times 45)$$

$$R_D = \frac{3000 + 1500 + 13\,500}{60} = 300\,kN$$

Moments about D:

$$\circlearrowright = \circlearrowleft$$

$$R_A \times 60 = (200 \times 45) + (50 \times 30) + ((10 \times 30) \times 15)$$

$$R_A = \frac{9000 + 1500 + 4500}{60} = 250\,kN$$

(b) The SF diagram can be determined by working from the left-hand end of the beam as follows:

SF at A $= R_A = 250\,kN$
SF at B $= 250 - 200 = 50\,kN$
SF at C $= 250 - 200 - 50 = $ zero kN

Between C and D the SF varies uniformly by $10\,kN/m$, i.e. the graph line will be straight and will have a uniform slope.

shear force at D $= 250 - 200 - 50 - (30 \times 10) = -300\,kN$

These SF values can now be used to draw the SF diagram (Figure 9.32).

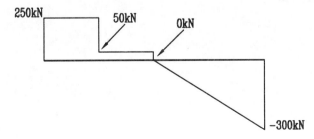

**Fig. 9.32** Shear force diagram.

(c) The BM diagram can be determined by first placing the hinge at B and working towards the right-hand end as follows:

hinge at $B = M_B = 250 \times 15 = 3750\,\text{kN}\,\text{m}$

hinge at $C = M_C = (250 \times 30) - (200 \times 15) = 4500\,\text{kN}\,\text{m}$

For the UDL take supplementary hinges at E and F:

$$\text{hinge at } E = M_E = (250 \times 40) - (200 \times 25) - (50 \times 10)$$
$$-\left((10 \times 10) \times \frac{10}{2}\right) = 4000\,\text{kN}\,\text{m}$$

$$\text{hinge at } F = M_F = (250 \times 50) - (200 \times 35) - (50 \times 20)$$
$$-\left((10 \times 20) \times \frac{20}{2}\right) = 2500\,\text{kN}\,\text{m}$$

$$\text{hinge at } D = M_D = (250 \times 60) - (200 \times 45) - (50 \times 30)$$
$$-\left((10 \times 30) \times \frac{30}{2}\right) = \text{zero}$$

These values can now be used to draw the BM diagram as shown in Figure 9.33.

**Example 9.12**

The loading system on the cantilever shown in Figure 9.34 comprises concentrated loads and a UDL. Construct:

(a) the SF diagram;
(b) the BM diagram.

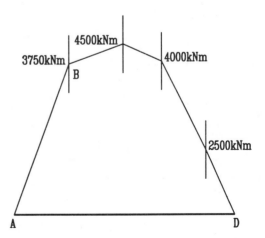

**Fig. 9.33** Bending moment diagram.

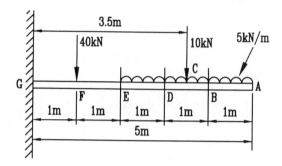

**Fig. 9.34** Cantilever beam.

*Solution*

(a) The SF diagram can be found by summing the vertical forces while working from the unsupported end towards the supported end as follows:

SF at A = zero

Shear force at C: note that the SF due to the UDL increases at a rate of 5 kN/m, so at C the SF is $(5 \times 1.5) = 7.5$ kN, due to the UDL only.

SF at C $= (5 \times 1.5) + 10 = 17.5$ kN
SF at E $= (5 \times 3) + 10 = 25$ kN
SF at F $= (5 \times 3) + 10 + 40 = 65$ kN
SF at G $= (5 \times 3) + 10 + 40 - 65 =$ zero kN

where $-65$ kN is the SF reaction at the support.

These values can now be used to construct the SF diagram shown in Figure 9.35.

(b) The BM diagram can be found by placing hinges at appropriate points along the beam while working from the unsupported end towards the built-in end. Convenient hinges should be placed at the point of

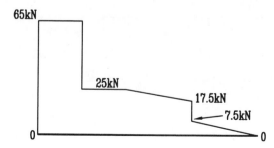

**Fig. 9.35** Shear force diagram.

application of the concentrated loads. In order to achieve more accuracy when drawing the BM diagram for the UDL, additional arbitrary points should be taken, say at B and D.

hinge at $A = M_A = $ zero kN m

hinge at $B = M_B = \left( (1 \times 5) \times \dfrac{1}{2} \right) = 2.5 \text{ kN m}$

hinge at $C = M_C = \left( (1.5 \times 5) \times \dfrac{1.5}{2} \right) = 5.625 \text{ kN m}$

hinge at $D = M_D = \left( (2 \times 5) \times \dfrac{2}{2} \right) + (10 \times 0.5) = 15 \text{ kN m}$

hinge at $E = M_E = \left( (3 \times 5) \times \dfrac{3}{2} \right) + (10 \times 1.5) = 37.5 \text{ kN m}$

hinge at $F = M_F = ((3 \times 5) \times 2.5) + (10 \times 2.5) = 62.5 \text{ kN m}$

hinge at $G = M_G = ((3 \times 5) \times 3.5) + (10 \times 3.5) + (40 \times 1) = 127.5 \text{ kN m}$

These values can now be used to construct the BM diagram shown in Figure 9.36.

## 9.9  SUMMARY

Figure 9.37 summarizes the sign conventions and SF and BM diagrams for standard support conditions for beams.

## 9.10  PROBLEMS

1. A horizontal beam 9.0 m long is simply supported at its ends and carries concentrated loads of 6.0, 2.0, 5.0 and 7.0 kN at distances of 2.0, 3.0, 5.0 and 8.0 m respectively from the left-hand end. Draw to scale the SF and BM diagrams, and state the magnitude and position of the maximum BM. Assume the mass of the beam to be negligible.

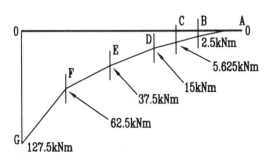

**Fig. 9.36** Bending moment diagram.

Sign conventions

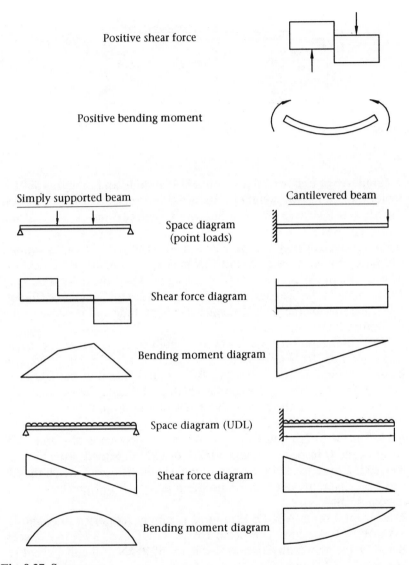

Positive shear force

Positive bending moment

Simply supported beam

Cantilevered beam

Space diagram
(point loads)

Shear force diagram

Bending moment diagram

Space diagram (UDL)

Shear force diagram

Bending moment diagram

**Fig. 9.37** Summary.

2. A beam ABCD is simply supported at A and C. AB = 5.0 m, BC = 5.0 m and CD = 3.0 m. The beam supports a UDL of 6.0 kN/m between B and D and concentrated loads of 25.0 kN at B and 10.0 kN at D. Neglecting the weight of the beam, draw to scale the SF and BM diagrams and determine the magnitude and position of the maximum BM.

3. Construct the SF and BM diagrams for the beam shown in Figure 9.38.

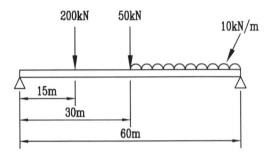

**Fig. 9.38**

4. A cantilever of uniform cross-section is 15 m long and of mass 1530 kg. In addition to its own weight it supports a central load of 14 kN. Sketch the SF and BM diagrams and find the magnitude and position of the maximum BM.

5. A uniform shaft ABCD, 4 m long and of mass 306 kg, is simply supported by bearings at A and C, which are 3 m apart. It carries two wheels, one of which has a mass of 408 kg at B and the other a mass of 102 kg at D. AD is 2 m and the overhang CD is 1 m. Sketch the SF and BM diagrams for the shaft and state the magnitude and position of the maximum BM.

6. A horizontal cantilever ABC, 4 m long, built-in at A, carries a concentrated load of 10 kN at its free end C. There is also a UDL of 4 kN/m between A and B; AB = 3 m and BC = 1 m. Neglecting the effects of gravity on the mass of the cantilever, draw to scale the SF and BM diagrams and state the maximum values of the SF and BM.

7. A horizontal beam ABCD, 13 m long, is simply supported at B and D. AB = 3 m, BC = 2 m and CD = 8 m. Concentrated loads of 4 and 7 kN act at A and C respectively, and a UDL of 1 kN/m extends from C to D. Neglecting the effects of gravity on the mass of the beam, sketch the SF and BM diagrams and determine the magnitude and position of the maximum BM.

8. A horizontal beam ABCD, 15 m long, is simply supported at A and C. AB = BC = CD = 5 m. The beam carries a UDL of 3 kN/m between B and D, together with concentrated loads of 10 kN at B and 4 kN at D. Sketch the BM and SF diagrams for the beam and determine the magnitude and position of the maximum BM. Assume the mass of the beam to be negligible.

# Stresses in beams and shafts  $\boxed{10}$

## 10.1 AIMS

- To introduce the neutral axis for a beam as an axis passing through the centroid of the section at which the bending stress is zero.
- To apply simple bending theory to a beam subjected to a bending moment.
- To evaluate the second moment of area for a beam section.
- To analyse the stresses in a shaft subjected to torsion.
- To evaluate the second polar moment of area for solid and hollow shafts of circular cross-section.

## 10.2 BEAMS AND SHAFTS

Chapter 8 introduced stresses and strains in components as a result of direct tensile, compressive and shear forces. This chapter takes that knowledge a stage further by considering the stresses resulting from either the bending of beams or torsion in shafts.

Types of beam and the loading of beams were discussed in Chapter 9. The significant feature of a beam is that it is long in relation to its depth. In Chapter 8, shear stress and shear strain were introduced by considering a horizontal bar fixed at one end. The point about the bar illustrated in Figure 8.16 was that it was short in relation to its depth. This means that it could not bend but deflected under the action of a shear force which caused a shear strain.

The difference between the deflection of a short bar and a long beam can be seen in Figure 10.1. Both are fixed at one end so that they act as cantilevers. In the case of the short bar, in Figure 10.1(a), the bar deflects through an angle $\phi$ that applies throughout its length. In the case of the beam, in Figure 10.1(b), the beam bends so that its final shape is a curve and the angle of the beam varies along its length. It is by studying the curvature of a beam that it is possible to analyse the stresses and this will be discussed later.

However, before doing so, it is necessary to emphasize that both beams and shafts are widely used in mechanical engineering, beams as structural components and shafts in drive systems.

At times, it is possible to have components that act as both beams and shafts. Figure 10.2 shows a geared shaft mounted on two bearings. If it is

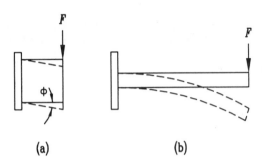

**Fig. 10.1** Bars deflecting as a result of shear force.

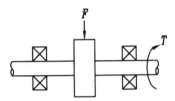

**Fig. 10.2** A geared shaft system.

assumed that the gear mates with another gear, on a parallel shaft, the shaft will be subjected to a torque because of the power being transmitted. In addition, the gear will be subjected to a side force, due to loading on the gear teeth. This side force will have a tendency to cause the shaft to bend and, therefore, act as a simply supported beam with the two bearings acting as the supports.

A similar model could be visualized for a shaft carrying a pulley, for which the tension in the belt round the pulley would again cause a significant side force. In this book, it is intended to analyse the bending of beams and torsion in shafts separately. The effect of combined bending and torsion is covered in more advanced texts. Nevertheless, the relationships developed within this chapter are very important since they are widely used in the analysis of mechanical components.

## 10.3 BENDING OF BEAMS

In Chapter 9 bending moments in beams were discussed. As the name implies, bending of a beam can be considered to occur as a result of moments applied to a beam, as illustrated in Figure 10.3.

The length of beam, shown in Figure 10.3, is in equilibrium because the two end moments are considered to be both equal and opposite. As a result of the moments, $M$, the beam will curve so that its material will be compressed on the top and stretched on the bottom. In other words,

**Fig. 10.3** Bending of a beam under the action of moments.

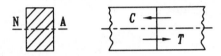

**Fig. 10.4** Position of the neutral axis.

the top part of the beam will be in compression and the bottom part in tension.

If the stresses in the beam vary from compressive on one side to tensile on the other, there must be a position where the stress, and the strain, are zero. This position is termed the **neutral axis**.

In the case of a beam having a rectangular cross-section, as shown in Figure 10.4, it can be shown that the neutral axis must lie on the horizontal centre-line of the section. The bending moment can be redefined in terms of two resultant forces, $C$ and $T$, either side of the neutral axis. Since the beam must be in equilibrium, it can be seen that the two resultant forces must be equal and opposite, and each acts over the same cross-sectional area. This can only be achieved if the neutral axis is situated on the centre-line.

Expressing this another way, the neutral axis passes through the **centroid**, i.e. the centre of area of the rectangular section. In fact, it can be shown that the neutral axis always passes through the centroid, irrespective of the cross-sectional shape of the beam.

For basic shapes, such as a rectangle, circle or triangle, the position of the centroid for each is well defined, as shown in Figure 10.5.

For more complex shapes than those shown in Figure 10.5, the position of the centroid can be found by taking moments of the individual areas contained within the overall shape. This can be illustrated by considering the general shape shown in Figure 10.6.

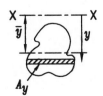

**Fig. 10.6** Moments of area about X–X.

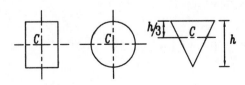

**Fig. 10.5** Position of centroids for basic shapes.

The moment of the area of the whole shape about axis XX, is

$$\bar{y} \times A$$

where $A$ is the total area of the shape. Now, this value must be equal to the moments of the individual areas making up the shape

$$\sum y \times A_y$$

summed for the whole shape. It follows that the distance from the axis XX to the centroid can be found by equating the moments:

$$\bar{y} \times A = \sum y A_y$$

and

$$\bar{y} = \frac{\sum y A_y}{A} \tag{10.1}$$

The application of equation (10.1) is illustrated in Example 10.1.

### Example 10.1

Determine the position of the centroid for the T-section defined in Figure 10.7.

*Solution*

The position of the centroid C can be found by considering the T-section to be made up of two constituent parts, A and B.
Applying equation (10.1):

$$\bar{y} = \frac{\sum y A_y}{A} = \frac{y_A A_A + y_B A_B}{A_A + A_B}$$

$$= \frac{5 \times (50 \times 10) + (10 + 25) \times (50 \times 10)}{(50 \times 10) + (50 \times 10)}$$

$$= \frac{40 \times (50 \times 10)}{2 \times (50 \times 10)} = 20\,\text{mm}$$

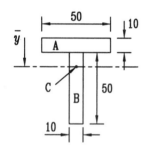

**Fig. 10.7**

### 10.4 THEORY OF BENDING

In order to analyse the bending of a beam, it is first necessary to make several simplifying assumptions, as follows:

1. The beam is bent only as a result of a bending moment.
2. The neutral axis passes through the centroid of the cross-sectional shape of the beam.

Both of these assumptions have been discussed in the preceding section. In addition, it is assumed that:

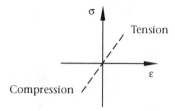

**Fig. 10.8** Stress–strain relationship.

3. The cross-sectional shape of the beam is symmetrical about the **plane of bending**. If the plane of bending is vertical, the types of shapes that are symmetrical about the plane include rectangular, circular, T- and I-sections.
4. The stress–strain relationship for the material remains within 'the elastic region and is the same in both tension and compression, as illustrated in Figure 10.8.
5. It is necessary to ensure that a plane, i.e. straight section across the beam, remains plane after bending. The significance of this is illustrated in Figure 10.9.

### 10.4.1 Stress due to bending

The significance of assumption (5) above can be seen from Figure 10.9. In the unbent condition, the two sections through the beams at AC and BD are plane and normal to the neutral axis. After bending, the sections remain plane and remain normal to the neutral axis. However, since the neutral axis has been bent, the two sections are no longer parallel but are at an angle of $\theta$ to each other. Before bending: AB = CD. After bending AB will be unchanged because it lies on the neutral axis, but CD will have increased to C′D′.

These lengths can be related to the radius of curvature, as follows:

$$AB = R\theta$$

and

$$C'D' = (R + y)\theta$$

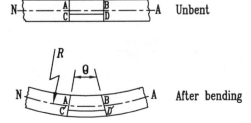

**Fig. 10.9** Change of shape due to bending.

The change in length between C'D' and CD is

$$C'D' - CD = C'D' - AB$$
$$= (R + y)\theta - R\theta = y\theta$$

This can be re-expressed in terms of the strain:

$$\varepsilon = \frac{\text{change in length}}{\text{original length}}$$

$$= \frac{y\theta}{R\theta} = \frac{y}{R}$$

From equation (8.3) the strain can be related to the stress in the beam and the modulus of elasticity for the material:

$$\varepsilon = \frac{\sigma}{E} = \frac{y}{R}$$

Rearranging gives the relationship

$$\frac{\sigma}{y} = \frac{E}{R} \tag{10.2}$$

Equation (10.2) is a useful means of analysing the stresses in a beam providing that the radius of curvature is known. There are some situations where the radius of curvature is known, as illustrated in Example 10.2 below.

Of more general relevance, equation (10.2) shows that the stress in the beam is a function of the distance from the neutral axis. A typical stress distribution for the beam considered above is presented in Figure 10.10. The maximum stress is at the section of the beam furthest from the neutral axis.

### Example 10.2

A steel strip, 50 mm wide and 2 mm thick, is bent around a cylindrical drum 1.5 m in diameter. Find the maximum stress in the strip if the value of $E$ is 200 GPa. The position of the neutral axis can be established by considering the cross-section of the strip (Figure 10.11).

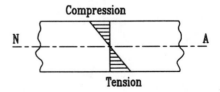

**Fig. 10.10** Stress distribution across a beam.

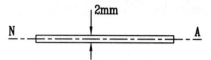

**Fig. 10.11**

*Solution*

The stress can be found using equation (10.2):

$$\frac{\sigma}{y} = \frac{E}{R}$$

which gives

$$\sigma = \frac{E}{R} y$$

Now $y = 1$ mm from Figure 10.11

$$\sigma = \frac{200 \times 10^9 \times 1}{0.75 \times 10^3} = 266.7 \times 10^6 \text{ Pa}$$

$$= 266.7 \text{ MPa}$$

*Note:* comparing this stress with the values quoted in Appendix A, this value is above the yield stress of mild steel.

In order to reduce the bending stress it is necessary to increase the radius of bending. This leads to the further assumption that: 'The radius of curvature of the bent beam must be large compared to the depth of the beam.'

### 10.4.2 Stress due to a bending moment

Although the relationship, defined in equation (10.2), for stress due to bending is valid, it is of limited application because very few bending situations have pre-defined radii of curvature. In practice, it is the applied bending moment that generally controls the stresses in a beam. It is, therefore, necessary to derive a relationship between stress and bending moment.

Consider the cross-section of a beam shown in Figure 10.12. For the element shown with area $\delta A$, at a distance $y$ from the neutral axis, the stress will be $\sigma$. It follows that the longitudinal force acting on the element is

$$F_y = \sigma \times \delta A$$

But the force, $F_y$, creates a moment about the neutral axis:

$$M_y = (\sigma \delta A) y \qquad (10.3)$$

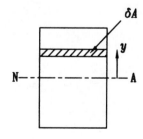

**Fig. 10.12** Cross-section of a beam.

From equation (10.2), it can be shown that

$$\sigma = \frac{E}{R} y$$

which, when substituted in equation (10.3), gives

$$M_y = \frac{E}{R}(\delta A y^2)$$

Now the total moment for the whole section is the sum of all the individual values of $M_y$:

$$M = \sum M_y = \frac{E}{R} \sum y^2 \delta A$$

The summation of $\sum y^2 \delta A$ for the whole section is termed the **second moment of area** for the cross-sectional shape about the neutral axis and is given the symbol $I$, so that

$$M = \frac{E}{R} I \qquad\qquad (10.4)$$

Combining equations (10.2) and (10.4) gives the complete relationship for the theory of bending:

$$\frac{M}{I} = \frac{\sigma}{y} = \frac{E}{R} \qquad\qquad (10.5)$$

Although the second moment of area, $I$, was introduced by considering a beam of rectangular cross-section, equation (10.5) is valid for all beam cross-sections providing that they are symmetrical about the plane of bending.

Elsewhere in the book, the symbol $I$ is used to denote the 'moment of inertia'. The use of the same symbol for two different variables should not cause confusion because the two are used in different applications. The moment of inertia is the second moment of mass with units of $\text{kg}\,\text{m}^2$, whereas the second moment of area has units of $\text{m}^4$.

### Example 10.3

A beam has a rectangular section 100 mm deep and a second moment of area, $I$, of $4 \times 10^{-6}\,\text{m}^4$ (Figure 10.13). Determine the bending moment that can be applied to the beam if the maximum operating stress is 50 MPa.

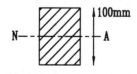

**Fig. 10.13**

*Solution*

The bending moment can be found using equation (10.5):

$$\frac{M}{I} = \frac{\sigma}{y} = \frac{E}{R}$$

which can be rearranged to give

$$M = \frac{\sigma}{y} I$$

From Figure 10.13, $y = 50\,\text{mm} = 0.05\,\text{m}$ and

$$M = \frac{50 \times 10^6 \times 4 \times 10^{-6}}{0.05} = 4000\,\text{N m}$$

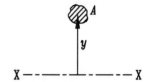

**Fig. 10.14** Position of an area with relation to an axis.

*Note:* if $E$ for the material is 200 GPa, it can be shown that the radius of curvature, $R$, is 200 m. For a depth of beam of 100 mm this means that $R/\text{depth} = 2000$.

## 10.5 SECOND MOMENT OF AREA

The 'second moment of area' is precisely what it says, the second moment of an area about a defined axis. Consider the small area shown in Figure 10.14. The second moment of area about axis X–X is

$$I_{\text{XX}} = y^2 A$$

where $y$ is the distance from the axis to the centroid of the area. It can be seen that the greater the distance from the axis, the larger the value of $I$.

Relating this to equation (10.5) for bending, the greater the value of $I$, the greater the stiffness of the beam because the radius of curvature is directly proportional to $I$:

$$R = \left(\frac{E}{M}\right) I$$

and the greater the radius of curvature, the less the deflection of the beam.

Comparing the three beam sections shown in Figure 10.15, all three have the same cross-sectional area but each has a different value of the second moment of area. By taking the rectangular section shown in Figure 10.15(a) and turning it on its edge, Figure 10.15(b), the value of the second moment of area is increased and the section provides a stiffer beam. By redistributing the material into the form of an I-section, as shown in Figure 10.15(c), the second moment of area is increased further because more of the material is at a greater distance from the neutral axis. This is why I-section beams are so widely used in practice. Data on

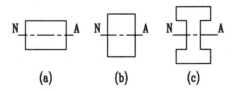

**Fig. 10.15** Comparisons of beam sections.

a range of I-section joists and beams can be found in Howatson, Lund and Todd (1991).

### 10.5.1 Second moment of area for a rectangle

To evaluate the values of the second moment of area for each of the sections shown in Figure 10.15, it is possible to apply one relationship:

$$I_{XX} = \frac{bd^3}{12} + Ah^2 \tag{10.6}$$

where $I_{XX}$ is the second moment of area about an axis parallel to, and at a distance $h$ from, the neutral axis. Equation (10.6) is valid for any rectangular section and $b$, $d$ and $A$ refer to the breadth, depth and cross-sectional area of the section respectively. A derivation of equation (10.6) is given in Appendix B.

Applying equation (10.6) to any section comprising several rectangular shapes depends on the position of the shapes with relation to the neutral axis of the section. In the case of the two rectangles shown in Figure 10.15(a) and (b), the neutral axis passes through the centroid of each and, applying equation (10.6), the $I_{NA}$ of each equals $I_{XX}$ for which the distance $h$ is zero. In other words, the second moment of area of a single rectangle about its neutral axis is

$$I_{NA} = \frac{bd^3}{12} + A(0)^2 = \frac{bd^3}{12}$$

In the case of an I-section, as shown in Figure 10.16, the complete section can be broken down into three separate rectangular shapes A, B and C, as shown. The second moment of area for each rectangle can then be found, using equation (10.6), and the total second moment of area for the section found by summing the individual values:

$$I_{section} = I_A + I_B + I_C$$

where each value of $I$ is found with respect to the neutral axis.

### Example 10.4

Determine the second moment of area for an I-section having a breadth of 80 mm, depth of 100 mm, flange thickness of 10 mm and web thickness of 8 mm (Figure 10.17).

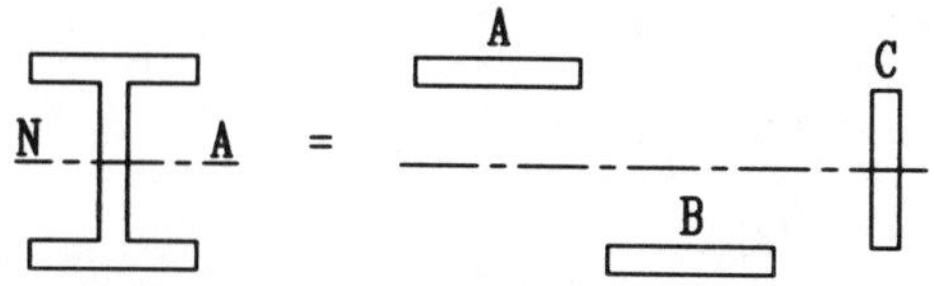

**Fig. 10.16** Subdividing an I-section into separate rectangles.

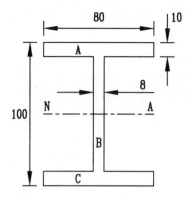

**Fig. 10.17**

*Solution*

The second moment of area can be found by applying equation (10.6) to the rectangles A, B and C as defined above.

For A:

$$I_A = \frac{bd^3}{12} + Ah^2$$

$$= \frac{80 \times (10)^3}{12} + (80 \times 10) \times (45)^2$$

$$= 1\,626\,667 \, \text{mm}^4 = 1.63 \times 10^{-6} \, \text{m}^4$$

For B:

$$I_B = \frac{bd^3}{12} + Ah^2$$

$$= \frac{8 \times (80)^3}{12} + (8 \times 80)(0)^2$$

$$= 341\,333 \, \text{mm}^4 = 0.34 \times 10^{-6} \, \text{m}^4$$

For C: from symmetry, $I_C = I_A$:

$$I_C = 1.63 \times 10^{-6} \, \text{m}^4$$

For the section:

$$I_{NA} = (1.63 + 0.34 + 1.63) \times 10^{-6}$$

$$= 3.6 \times 10^{-6} \, \text{m}^4$$

## Example 10.5

A simply supported beam of 4 m span carries a central load of 1 kN. If the beam has a T-section as shown in Figure 10.18, calculate the maximum tensile and compressive stresses in the beam.

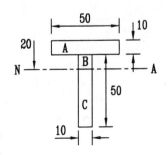

**Fig. 10.18**

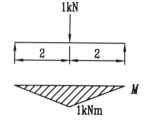

1kNm

**Fig. 10.19**

*Note:* from Example 10.1, the centroid of this T-section is situated at a distance of 20 mm from the top edge.

*Solution*

The stresses can be found using equation (10.5):

$$\frac{M}{I} = \frac{\sigma}{y} = \frac{E}{R}$$

and

$$\sigma = \frac{M}{I}y$$

However, to evaluate the stresses, it is first necessary to find the maximum bending moment and the value of $I$ for the section. From the discussion in Chapter 9 the bending moment diagram is as shown in Figure 10.19.

The second moment of area can be found by applying equation (10.6) to the rectangles A, B and C as defined above.

For A:

$$I_A = \frac{bd^3}{12} + Ah^2$$

$$= \frac{50 \times (10)^3}{12} + (50 \times 10) \times (15)^2$$

$$= 116\,667\,\text{mm}^4 = 0.117 \times 10^{-6}\,\text{m}^4$$

For B:

$$I_B = \frac{bd^3}{12} + Ah^2$$

$$= \frac{10 \times (10)^3}{12} + (10 \times 10) \times (5)^2$$

$$= 3333\,\text{mm}^4 = 0.003 \times 10^{-6}\,\text{m}^4$$

For C:

$$I_C = \frac{bd^3}{12} + Ah^2$$

$$= \frac{10 \times (40)^3}{12} + (10 \times 40) \times (20)^2$$

$$= 213\,333\,\text{mm}^4 = 0.213 \times 10^{-6}\,\text{m}^4$$

For the section:

$$I_{NA} = (0.117 + 0.003 + 0.213) \times 10^{-6}$$

$$= 0.333 \times 10^{-6}\,\text{m}^4$$

Substituting in equation (10.5):

$$\sigma = \frac{M}{I}y$$

$$= \frac{1000 \times y}{0.333 \times 10^{-6}}$$

At the top edge, $y = 20\,$mm and the material will be in compression:

$$\sigma_C = \frac{1000 \times 0.02}{0.333 \times 10^{-6}} = 60.1 \times 10^6\,\text{Pa}$$

$$= 60.1\,\text{MPa}$$

At the bottom edge, $y = 40\,$mm and the material will be in tension:

$$\sigma_T = \frac{1000 \times 0.04}{0.333 \times 10^{-6}} = 120.1 \times 10^6\,\text{Pa}$$

$$= 120.1\,\text{MPa}$$

### 10.5.2  Second moment of area for a circle

Just as a rectangle is a useful building block in the analysis of a wide range of beam sections, so a similar approach can be used for circular sections. The second moment of area of a circle can be evaluated using the relationship

$$I_{xx} = \frac{\pi r^4}{4} + Ah^2 \tag{10.7}$$

where $I_{xx}$ is the second moment of area for an axis parallel to, and at a distance $h$ from, the neutral axis, and $r$ is the radius of the circular section.

### Example 10.6

A circular tubular beam, 60 mm outside diameter and 50 mm inside diameter, is simply supported over a span of 3 m and carries a uniformly distributed load of 1 kN/m (Figure 10.20). Calculate the maximum stress in the tube.

*Solution*

The stress can be found using equation (10.5):

$$\sigma = \frac{M}{I}y$$

The maximum bending moment is in the centre of the span, as shown by Figure 10.21. Reaction at the left-hand side is 1.5 kN and at the centre of

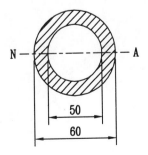

**Fig. 10.20**

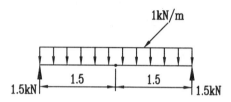

**Fig. 10.21**

the span:

$$M = (1.5 \times 1.5) - (1.5 \times 1) \times 0.75$$

$$= 1.125\,\text{kN}\,\text{m} = 1125\,\text{N}\,\text{m}$$

The second moment of area can be found by applying equation (10.7) to the tubular section:

$$I_{NA} = I_{outer} - I_{inner}$$

$$= \frac{\pi(30)^4}{4} + A_o(0)^2 - \frac{\pi \times (25)^4}{4} + A_i(0)^2$$

$$= \frac{\pi}{4}(30^4 - 25^4) = 329\,376\,\text{mm}^4$$

$$= 0.329 \times 10^{-6}\,\text{m}^4$$

Substituting in equation (10.5):

$$\sigma = \frac{M}{I}y = \frac{M}{I}r$$

$$= \frac{1125 \times 0.03}{0.329 \times 10^{-6}} = 102.6 \times 10^6\,\text{Pa}$$

$$= 102.6\,\text{MPa}$$

## 10.6 STRESSES IN CIRCULAR SHAFTS

When a solid circular shaft is subjected to an applied torque, as shown in Figure 10.22, it will twist. The amount of twist determines the shear strain in the shaft and, as a consequence, the magnitude of the shear stress. The twist resulting from the applied torque is termed 'torsion'.

Consider a straight line, AB, scribed along the external surface of a shaft as shown in Figure 10.22. Under the action of a torque $T$, the shaft will twist so that point B moves to position B′ and the line becomes AB′.

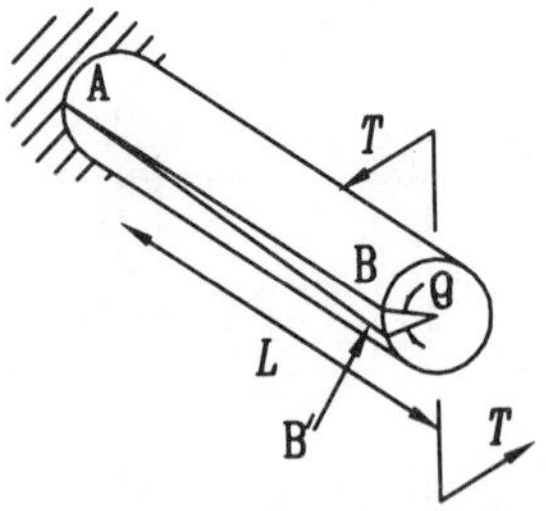

### 10.6.1 Stress due to torsion

**Fig. 10.22** Shaft subjected to torsion.

From the definition of shear strain, in section 8.7.1, the movement BB′ can be equated to the length of the shaft through which the torsion is taking place:

$$\phi = \frac{BB'}{L}$$

About the centre of the shaft the movement BB' can be related to an angle of t†wist:

$$\theta = \frac{BB'}{r}$$

so that

$$\phi L = \theta r$$

Rearranging

$$\phi = \frac{\theta r}{L}$$

However, from the definition of shear modulus:

$$\phi = \frac{\tau}{G}$$

and

$$\frac{\tau}{G} = \frac{\theta r}{L}$$

This is generally presented in the form

$$\frac{\tau}{r} = \frac{G\theta}{L} \tag{10.8}$$

It will be noticed that the form of equation (10.8) is very similar to that of equation (10.2) relating to beams. It also indicates that the shaft will have a smaller angle of twist, the greater the value of the shear modulus.

### Example 10.7

A steel shaft of 50 mm diameter and 800 mm long is subjected to torsion. If the shear stress is to be limited to 50 MPa, determine the maximum angle of twist in the shaft. Take the shear modulus as 80 GPa (see Figure 10.22).

*Solution*

The angle of twist can be found using equation (10.8):

$$\frac{\tau}{r} = \frac{G\theta}{L}$$

Rearranging:

$$\theta = \frac{\tau L}{Gr}$$

$$= \frac{50 \times 10^6 \times 0.8}{80 \times 10^4 \times 0.025} = 0.02\,\text{rad}$$

*Note:* the angle in radians can be converted to degrees by recognizing that there are $2\pi$ rad in a full circle of $360°$:

$$\theta = \frac{0.02 \times 360}{2\pi} = 1.15°$$

### 10.6.2 Stress due to a torque

Although equation (10.8) is valid for shafts subjected to torsion, it is of limited application because, for most torsion situations, it is the torque applied to the shaft that determines the amout of twist and the resulting shear stresses. It is, therefore, necessary to derive a relationship between shear stress and torque.

Consider the cross-section of a circular shaft shown in Figure 10.23. For the elemental ring shown at radius $r$, the area is

$$\text{elemental area} = 2\pi \times r \times \delta r$$

Since the shear stress acting at radius $r$ is $\tau$, the shear force acting on the elemental ring is

$$F_s = \tau(2\pi r \delta r)$$

The shear force creates a moment about the centre-line of the shaft:

$$M_r = F_s \times r = \tau(2\pi r^2 \delta r)$$

Summing the moments for all the elemental rings in the shaft gives the torque for the whole shaft:

$$T = \sum M_r = \tau \sum 2\pi r^3 \delta r$$

From equation (10.8) the shear stress can be expressed as

$$\tau = \frac{G\theta}{L} r$$

and

$$T = \frac{G\theta}{L} r \sum 2\pi r^3 \delta r$$

$$= \frac{G\theta}{L} \sum 2\pi r^3 \delta r$$

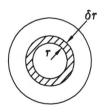

**Fig. 10.23** Cross-section of a shaft.

The summation of $\sum 2\pi r^3 \delta r$ for the whole shaft is termed the **second polar moment of area** for the circular shaft about the longitudinal centre-line, and is given the symbol $J$, so that

$$T = \frac{G\theta}{L}J \tag{10.9}$$

Combining equations (10.8) and (10.9) gives the complete relationship for torsion within circular shafts:

$$\frac{T}{J} = \frac{\tau}{r} = \frac{G\theta}{L} \tag{10.10}$$

It will be noticed that the form of this equation is similar to that given earlier in equation (10.5) for the bending of beams.

In order to apply equation (10.10), it is necessary to evaluate the torque, $T$, and the second polar moment of area, $J$. In Chapter 5, it was shown that the power transmitted by a shaft can be related to the torque by

$$W = \omega T$$

where $\omega$ is the angular velocity of the shaft in rad/s. The second polar moment of area for a circular shaft is

$$J = \frac{\pi}{2}r^4 \tag{10.11}$$

Clearly, the value of the second polar moment of area increases rapidly as the radius increases. This means that most of the torque being transmitted by the shaft is being carried by the outer portion at a radius furthest from the centre-line. It follows that, where weight is important, the shaft can be made hollow with little effect on the overall power that can be transmitted. The application of equation (10.10) to a hollow shaft is demonstrated in Example 10.9 below.

**Example 10.8**

Find the power that can be transmitted by the 50 mm diameter shaft defined in Example 10.7 if it rotates at 1200 rev/min.

*Solution*

The torque transmitted by the shaft can be found using equation (10.10):

$$\frac{T}{J} = \frac{\tau}{r} = \frac{G\theta}{L}$$

Rearranging

$$T = \frac{\tau J}{r}$$

Using equation (10.11)

$$J = \frac{\pi}{2}r^4 = \frac{\pi}{2}(25)^4 = 613\,592.3\,\text{mm}^4$$

$$= 6.13 \times 10^{-7}\,\text{m}^4$$

The maximum shear stress was defined as 50 MPa in Example 10.7:

$$T = \frac{50 \times 10^6 \times 6.13 \times 10^{-7}}{0.025} = 1227.2\,\text{N m}$$

The angular velocity of the shaft is

$$\omega = \frac{2\pi \times 1200}{60} = 125.67\,\text{rad/s}$$

and the power transmitted is

$$W = \omega T = 125.67 \times 1227.2$$
$$= 154\,214.5\,\text{W}$$
$$= 154.2\,\text{kW}$$

**Example 10.9**

Determine the power that can be transmitted by a hollow shaft of 40 mm outside diameter and 20 mm inside diameter, rotating at 2400 rev/min if the maximum shear stress is 40 MPa (Figure 10.24).

*Solution*

The second polar moment of area for the hollow shaft can be found from $J$ for the outer radius less $J$ for the hollow. From equation (10.11):

$$J = \frac{\pi}{2}(20)^4 - \frac{\pi}{2}(10)^4 = 235\,619.5\,\text{mm}^4$$

$$= 2.356 \times 10^{-7}\,\text{m}^4$$

The torque can be found using equation (10.10):

$$T = \frac{\tau J}{r} = \frac{40 \times 10^6 \times 2.356 \times 10^{-7}}{0.02} = 471.2\,\text{N m}$$

The angular velocity of the shaft is

$$\omega = \frac{2\pi \times 2400}{60} = 251.33\,\text{rad/s}$$

and the power transmitted is

$$W = \omega T = 251.33 \times 471.2$$
$$= 118\,425.5\,\text{W}$$
$$= 118.4\,\text{kW}$$

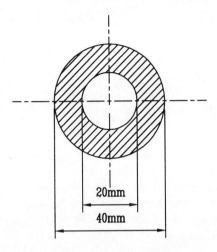

**Fig. 10.24**

*Note:* if the shaft had been solid with the same outside diameter, the power transmitted would be increased to 126.3 kW, an increase of just under 7%, confirming that most of the power is transmitted by the outer portion of the shaft.

## 10.7 SUMMARY

Key equations that have been introduced in this chapter are as follows.
  Position of the centroid of a shape about an axis:

$$\bar{y} = \frac{\sum y A_y}{A} \tag{10.1}$$

Bending of beams:

$$\frac{M}{I} = \frac{\sigma}{y} = \frac{E}{R} \tag{10.5}$$

Second moment of area for a rectangular section:

$$I_{xx} = \frac{bd^3}{12} + Ah^2 \tag{10.6}$$

Second moment of area for a circular section:

$$I_{xx} = \frac{\pi r^4}{4} + Ah^2 \tag{10.7}$$

Torsion in circular shafts:

$$\frac{T}{J} = \frac{\tau}{r} = \frac{G\theta}{L} \tag{10.10}$$

Second polar moment of area for a circular shaft:

$$J = \frac{\pi}{2} r^4 \tag{10.11}$$

## 10.8 PROBLEMS

### 10.8.1 Bending of beams

1. A rectangular beam 50 mm wide and 120 mm deep is simply supported over a span of 5 m. If the beam supports a single load of 1 kN at the mid span, determine the maximum bending stress.
2. A steel wire of 2 mm diameter is bent around a drum of 600 mm radius. If the modulus of elasticity for the steel is 200 GPa, find the maximum bending stress in the wire.
3. For the situation defined in problem 2, determine the bending moment applied to the wire.
4. A hollow tube of 50 mm external diameter and 40 mm internal diameter is simply supported over a span of 4 m. If the tube carries a uniformly distributed load of 1 kN/m along its length, determine the maximum bending stress.
5. An I-section beam is simply supported over a span of 5 m and carries a uniformly distributed load of 4 kN/m. The section is 200 mm deep and 100 mm wide overall. The top and bottom flanges are 20 mm thick. The web is 10 mm thick. Calculate the maximum bending stress in the beam.
6. A simply supported rectangular beam has a span of 4 m and carries a load of 6 kN at mid span. If the depth of the section is 200 mm, determine the required width of the section if the maximum bending stress is to be limited to 20 MPa.

### 10.8.2 Torsion in shafts

*Note:* for the following problems, take $G$ to be 80 GPa.

7. A solid steel shaft has a diameter of 50 mm. If it is subjected to a torque of 500 N m, calculate:
   (a) the maximum shear stress;
   (b) the angular twist for a unit length of shaft.
8. A solid steel shaft of 80 mm diameter is required to transmit 75 kW at 240 rev/min. Determine the maximum shear stress in the shaft.
9. A hollow propeller shaft on a shaft has an outside diameter of 300 mm and an inside diameter of 200 mm. If it rotates at a constant speed of 120 rev/min, what power can be transmitted. Assume a maximum shear stress of 40 MPa.
10. A solid steel shaft is required to transmit 100 kW at 1800 rev/min. The shear stress in the shaft must not exceed 50 MPa and the twist must not exceed 1° per unit length of shaft. Find the minimum diameter of shaft to satisfy these conditions.

# Thermofluid situations $\boxed{11}$

## 11.1 AIMS

- To introduce the types of fluid used in thermofluid situations.
- To define the hydrostatic pressure for a fluid having a constant density.
- To discuss pressure-measuring devices such as the barometer and manometer.
- To introduce the non-flow energy equation for a closed thermofluid system undergoing a process.
- To introduce the first law of thermodynamics for a closed system undergoing a cycle.

## 11.2 THERMOFLUID MECHANICS

In Chapter 1 a thermofluid system was introduced as a means of modelling a situation. The basic features of such a system were defined in Figure 1.5, which illustrated the thermofluid system necessary to model gas in the cylinder of a reciprocating engine. Such a system is defined as a 'closed' system because the amount of gas in the cylinder is fixed under the conditions defined.

Where a fluid can flow into, and out of, a system, the type of system is termed an 'open' system. Both closed and open systems are considered in more detail later in the text but, before doing so it is necessary to consider what is meant by the term 'thermofluid' in this context.

Thermofluid relates to situations that can be modelled using a thermofluid system and analysed by means of 'thermofluid mechanics'. Thermofluid mechanics combines the two disciplines of thermodynamics and fluid mechanics.

Thermodynamics is largely concerned with the conversion of heat into work within engines. What is generally referred to as 'heat' is, in fact, thermal energy. Heat is one form of energy that can cross the boundary of a thermofluid system. Thermal energy, on the other hand, is released by either burning fossil fuels or within nuclear reactions. Some part of the thermal energy may be transferred as heat in order to be converted into work. The work may be in the form of mechanical power from the output shaft of an engine or in the form of a propulsive force produced by an aircraft engine.

Since heat is transferred as a result of a temperature difference, it is essential to have some understanding of heat transfer, and this topic is discussed in Chapter 15.

In order to analyse thermodynamic situations it is also essential to have a knowledge of the properties and behaviour of fluids, hence the inclusion of fluid mechanics with thermodynamics to form an integrated study of thermofluid mechanics.

## 11.3  FLUIDS

A **fluid** is defined as a substance that can flow. As such, the term embraces any or all of the following physical states: gases, liquids and vapours. In addition, the sudy of fluid behaviour is based on the assumption that the chemical composition of a fluid in any situation remains constant. In other words, a fluid is a pure substance.

This does not restrict the study of fluids only to those comprising single elements. A mixture of various elements or compounds can be considered as a pure substance, providing that the overall composition does not change. Air is a fluid that is widely used in thermofluid situations. As such, it is taken as a pure substance, even though it is made up of several gases including oxygen, nitrogen and carbon dioxide. It is assumed that none of the constituent gases undergoes a chemical change during a thermofluid process.

Fluids tend to be referred to by the physical state in which they exist under normal atmospheric conditions. This avoids any confusion when considering a fluid that may exist in all the different liquid, vapour and gaseous states. For example, water exists as a liquid under normal atmospheric conditions. Raising its temperature to the boiling point causes it to change to steam, a vapour. When all the water has been changed to steam, additional heat input causes the temperature to rise above the boiling point into what is termed the superheated region. If the temperature of the superheated steam is sufficiently high, then it achieves what is effectively a gaseous state.

This can be visualized by considering oxygen, a gas under normal atmospheric conditions. If the temperature of oxygen is reduced to $-183\,°C$, it changes from a gas to a vapour and then, eventually, to a liquid.

Referring to fluids by their physical state under normal atmospheric conditions means that air, oxygen and hydrogen, for example, are considered as gases, while water is considered to be a liquid. This enables fluids to be classified, but in order to analyse their behaviour, it is necessary to define the state of a particular fluid in terms of its physical properties.

There are six thermofluid properties that define the state of the fluid, any of which may be necessary for a given situation:

1. the measurable properties:
   (a) temperature, $T$
   (b) pressure, $P$
2. the mechanical properties of the fluid:
   (a) density, $\rho$
   (b) viscosity, $\mu$

Single start　　Two-start　　Four-start

# Properties of fluids $\boxed{12}$

## 12.1   AIMS

- To define the equation of state for an ideal gas.
- To define the specific heats for an ideal gas and the relationship between them.
- To define the basic relationships for an ideal gas undergoing an adiabatic process.
- To explain the changes that take place during a two-phase process.
- To introduce the saturated conditions, dryness fraction and super-heated conditions for a two-phase fluid.

## 12.2   PROPERTIES TO BE CONSIDERED

Fluids and their properties have already been introduced in section 11.3. Fluids are defined by their physical state as: gases, liquids or vapours, and these defined states form the basis for the present chapter.

Although the title of this chapter implies a study of all the properties necessary to define the state of a fluid, in reality it is proposed to concentrate on those properties necessary to analyse processes within a closed thermofluid system. These properties are:

- temperature, $T$ (K)
- pressure, $P$ (Pa)
- density, $\rho$ (kg/m$^3$)
- internal energy, $u$ (J/kg)
- enthalpy, $h$ (J/kg).

Viscosity is defined later in Chapter 14 in connection with flow of fluids through pipes and systems having friction.

Since the study of these properties depends on the fluids being studied, it is proposed to consider the properties of gases separately from the properties of liquids and vapours.

## 12.3   IDEAL GASES

In order to simplify the analysis of gases, it is necessary to introduce the concept of an 'ideal gas'. Just as real situations have to be modelled in order for them to be analysed, so real gases are assumed to behave as ideal gases. Within reasonable limits this assumption is realistic and most gases behave close to the ideal within the range of temperatures and pressures found in most thermofluid situations.

The concept of an ideal gas is one in which the physical matter within the gas is assumed to be in the form of spheres of negligible volume. The spheres move at random and undergo perfectly elastic collisions with the walls of the container and with each other.

This perfect behaviour means that the properties of the ideal gas can be defined precisely by means of the gas laws.

### 12.3.1 Gas laws

Behaviour of an ideal gas can be defined by means of the two gas laws: Boyle's law and Charles' law.

Robert Boyle developed his famous law for an 'elastic' fluid, i.e. a gas, in 1662. It states that for a quantity of gas at constant temperature, the variation of pressure with volume follows the relationship:

$$P \times V = \text{constant} \tag{12.1}$$

As an engineering aside, this relationship was quoted by James Watt a century later in connection with his patent for an expansion steam engine.

The second gas law was developed by Jacques Charles in 1787. It states that for a quantity of gas at constant pressure, the variation of volume with temperature follows the relationship:

$$\frac{V}{T} = \text{constant} \tag{12.2}$$

where $T$ is the absolute temperature of the gas.

There are many situations which do not take place at either constant temperature or constant pressure. For such situations it is possible to combine the two gas laws to give a more general relationship.

For a given quantity, or mass, of gas, the volume can be related to both the pressure and temperature. From equation (12.1):

$$V = \text{constant} \times \frac{1}{P}$$

and from equation (12.2):

$$V = \text{constant} \times T$$

Combining gives

$$V = \text{constant} \times \frac{T}{P}$$

which can be rearranged as

$$\frac{PV}{T} = \text{constant} \tag{12.3}$$

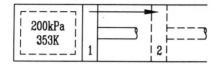

**Fig. 12.1**

One definition of an ideal gas is that it obeys the relationship defined in equation (12.3).

**Example 12.1**

An ideal gas is contained in a piston and cylinder assembly, with an initial pressure of 200 kPa and temperature of 80 °C (Figure 12.1). If the piston moves so that the volume is doubled and the temperature drops to 20 °C, find the final pressure of the gas.

*Solution*

The final pressure can be found by means of equation (12.3):

$$\frac{PV}{T} = \text{constant}$$

Therefore

$$\frac{P_1 V_1}{T_1} = \frac{P_2 V_2}{T_2}$$

Now

$$P_1 = 200 \, \text{kPa}$$
$$T_1 = 80 + 273 = 353 \, \text{K}$$
$$T_2 = 20 \times 273 = 293 \, \text{K}$$
$$V_2 = 2 \times V_1$$

Substituting gives

$$\frac{200 \times 10^3 \times V_1}{353} = \frac{P_2 \times 2 V_1}{293}$$

and

$$P_2 = \frac{200 \times 10^3 \times 293}{2 \times 353} = 83\,003 \, \text{Pa} = 83 \, \text{kPa}$$

*Note:* equation (12.3) is only valid if the values of $T$ are in absolute temperature, i.e. K.

### 12.3.2 Equation of state

Although equation (12.3) is valid for an ideal gas, it is limited to predicting the change in properties due to a process. In order to assess the state of an ideal gas under equilibrium conditions, it is necessary to put a value to the constant represented by $(PV/T)$. This can be done by creating an 'equation of state' for an ideal gas.

Equation (12.3) is expressed in the form

$$\frac{PV}{T} = \text{constant}$$

However, the 'constant' can only be constant for a particular quantity, or mass, of gas because the volume depends on the mass.

To make equation (12.3) independent of mass, it is necessary to divide by the actual mass of the gas, $m$:

$$\frac{PV}{Tm} = \text{constant}$$

However, $m/V$ is an expression for density, $\rho$, so that

$$\frac{P}{\rho T} = \text{constant}$$

The constant is given the symbol $R$ and is generally referred to as the 'gas constant'. Using the gas constant the equation of state can be expressed as

$$P = \rho R T \qquad\qquad\qquad (12.4)$$

By substituting the units of the variables into equation (12.4), it is possible to determine the units of the gas constant:

$$R = \frac{P}{\rho T}$$

$$= \frac{(\text{Pa})}{(\text{kg/m}^3)(\text{K})} = \frac{(\text{N/m}^2)}{(\text{kg/m}^3)(\text{K})} = \frac{\text{N m}}{\text{kg K}} = \text{J/kg K}$$

The value of the gas constant depends on the particular gas being considered. A universal gas constant that applies to all ideal gases is discussed in the next section.

In passing, it should be noted that since an ideal gas is one that obeys equation (12.3), it must also obey the equation of state (12.4). In fact, by using this as a basis, it is possible to define an ideal gas as one that obeys the equation of state.

### Example 12.2

A room, $2 \times 4 \times 3\,\text{m}$ in size, contains air at standard atmospheric conditions of $15\,°\text{C}$ and $101.33\,\text{kPa}$. If the gas constant for air is $0.287\,\text{kJ/kg K}$, find the mass of air in the room.

*Solution*

The mass of air can be calculated by first determining the air density from equation (12.4):

$$P = \rho RT$$

so that

$$\rho = \frac{P}{RT}$$

Now $T = 15 + 273 = 288\,\text{K}$ and

$$\rho = \frac{101.33 \times 10^3}{0.287 \times 10^3 \times 288} = 1.226\,\text{kg/m}^3$$

From the definition of density:

$$\rho = \frac{m}{V}$$

so that

$$m = \rho \times V$$
$$= 1.226 \times (2 \times 4 \times 3) = 29.42\,\text{kg}$$

### 12.3.3 Molecular unit

The gas constant defined in equation (12.4) is based on a unit mass of the gas being considered. By relating the gas constant to a fixed number of molecules in a gas, a universal gas constant can be established. The problem then becomes one of equating the actual mass of the gas, in kilograms, to the number of molecules. This is achieved by means of a molecular unit of mass, termed the 'kilogram mole'.

Avogadro's hypothesis states that a given volume of gas, at constant temperature and pressure, will contain a fixed number of molecules irrespective of the particular gas being considered. It follows that the actual mass of gas will depend on its **molecular weight**. The molecular weight of a substance is defined as the ratio of the mass of a molecule of the substance compared with that of one atom of hydrogen. It is sometimes also referred to as the 'molar mass'. A brief list of some molecular weights is given in Table 12.1.

**Table 12.1** Molecular weights of some common substances

| Substance | Molecular symbol | Molecular weight |
|---|---|---|
| Carbon-12 | C | 12 |
| Helium | He | 4 |
| Hydrogen | $H_2$ | 2 |
| Nitrogen | $N_2$ | 28 |
| Oxygen | $O_2$ | 32 |
| Sulphur | S | 32 |

The molecular weight of oxygen is 32 and that for hydrogen is 2. A volume that contains 2 kg of hydrogen would contain 32 kg of oxygen, because they both would have the same number of molecules. The number of molecules in this instance is defined by the kilogram mole, which is given the symbol 'kmol'.

A kmol is a quantity that contains the same number of molecules as there are atoms in 12 kg of carbon-12. This quantity of molecules can be visualized as having a mass, but the mass depends on the molecular weight and can be calculated from

$$m = n \times M \qquad (12.5)$$

where $m$ is the mass of gas (kg), $n$ the number of kmol and $M$ the molecular weight of the gas. It follows from equation (12.5) that the units of molecular weight are

kg/kmol

Using the kmol as a unit of mass the gas constants can be related to one universal gas constant, $R_0$. The universal gas constant has the value of

$$R_0 = 8.314 \, \text{kJ/kmol K}$$

Since 1 kmol of a gas contains $M$ kilograms, the gas constant for any particular gas can be found by using the universal gas constant in the following relationship:

$$R = \frac{R_0}{M} \qquad (12.6)$$

### Example 12.3

By volume, 1 kmol of air contains 79% of nitrogen and 21% of oxygen. Calculate the mass of air contained.

*Solution*

A volume of a gas contains the same number of molecules irrespective of whether the gas is nitrogen, oxygen or a mixture of the two. Therefore

$$1 \, \text{kmol air} = 0.79 \, \text{kmol N}_2 + 0.21 \, \text{kmol O}_2$$

The mass of air can be found using equation (12.5):

$$m = (nM)\text{N}_2 + (nM)\text{O}_2$$
$$= 0.79 \times 28 + 0.21 \times 32$$
$$= 22.12 + 6.72 = 28.84 \, \text{kg}$$

### Example 12.4

Find the density of carbon dioxide ($CO_2$) at a pressure of 250 kPa and a temperature of 20 °C.

*Solution*

The molecular weight of $CO_2$ is

$$M(CO_2) = 12 + 32 = 44 \, \text{kg/kmol}$$

The gas constant can be found using equation (12.6):

$$R = \frac{R_0}{M} = \frac{8.314}{44} = 0.189 \, \text{kJ/kg K}$$

Using this value in the equation of state (12.4):

$$P = \rho R T$$

where $T = 20 + 273 = 293 \, \text{K}$,

$$\rho = \frac{P}{RT} = \frac{250 \times 10^3}{0.189 \times 10^3 \times 293} = 4.516 \, \text{kg/m}^3$$

## 12.4 SPECIFIC HEATS OF GASES

The specific heat of a substance is defined as the heat required to raise one unit mass of that substance through a temperature rise of one degree. In the SI system the specific heat has units of

$$J/\text{kg K}$$

Solids and liquids are considered to be incompressible, as the density does not change significantly, and the specific heat for such substances can be taken as constant.

In the case of a gas it is necessary to define two values of specific heat:

1. specific heat at constant volume, $C_V$;
2. specific heat at constant pressure, $C_P$.

These can be evaluated by considering both a constant volume process and a constant pressure process in a closed thermofluid system.

### 12.4.1 Constant volume process for an ideal gas

Consider a closed thermofluid system having a fixed boundary and containing 1 kg of an ideal gas, as shown in Figure 12.2. Applying the non-flow energy equation to the closed thermofluid system, equation (11.6) gives

$$q - w = u_2 - u_1$$

However, since the boundary is fixed, there is no movement of the boundary and the work done is zero:

$$q = u_2 - u_1$$

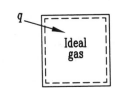

**Fig. 12.2** A closed system having constant volume.

Since the heat transferred from the surroundings takes place at constant volume, it follows that

$$q = C_V(T_2 - T_1) \tag{12.7}$$

It is possible to relate the specific heat at constant volume to the change in internal energy and the change in temperature:

$$C_V(T_2 - T_1) = u_2 - u_1$$

and for small changes:

$$C_V \Delta T = \Delta u$$

so that

$$C_V = \frac{\Delta u}{\Delta T} \tag{12.8}$$

This is true for all ideal gases irrespective of whether the process is at constant volume or not.

### Example 12.5

A closed thermofluid system contains 1 kg of an ideal gas. During a constant volume process 60 kJ of heat is transferred, causing the temperature of the gas to increase from 22 to 104 °C. Calculate the value of $C_V$ and the ratio of the final pressure to the initial pressure of the gas. (See Figure 12.2.)

*Solution*

The value of $C_V$ can be found using equation (12.7):

$$q = C_V(T_2 - T_1)$$

For 1 kg of gas, $q = 60$ kJ/kg and

$$(T_2 - T_1) = (104 - 22) = 82 \text{ K}$$

Therefore

$$C_V = \frac{60}{82} = 0.732 \text{ kJ/kg K}$$

From the equation of state (12.4):

$$P = \rho R T$$

Since the mass and volume do not change, $\rho$ must be constant, so that

$$\frac{P}{T} = \text{constant}$$

For the constant volume process:

$$\frac{P_1}{T_1} = \frac{P_2}{T_2}$$

Therefore

$$\frac{P_2}{P_1} = \frac{T_2}{T_1} = \frac{(104 + 273)}{(22 + 273)} = 1.278$$

### 12.4.2 Constant pressure process for an ideal gas

Consider a closed thermofluid system containing 1 kg of an ideal gas and undergoing a constant pressure process, as shown in Figure 12.3.

From the discussion in section 11.6.3, it was shown that during a constant pressure process, the heat transferred to a closed system is equal to the change of enthalpy of the fluid within the system. From equation (11.8):

$$q = h_2 - h_1$$

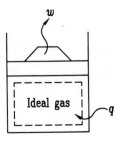

**Fig. 12.3** A closed system under constant pressure.

Since the heat from the surroundings is transferred at constant pressure, it follows that

$$q = C_P(T_2 - T_1) \tag{12.9}$$

Equating the change of enthalpy with the heat transferred, as defined in equation (12.9):

$$C_P(T_2 - T_1) = h_2 - h_1$$

and for small changes:

$$C_P \Delta T = \Delta h$$

so that

$$C_P = \frac{\Delta h}{\Delta T} \tag{12.10}$$

This definition of $C_P$ is true for all ideal gases irrespective of whether the process is at constant pressure or not.

### Example 12.6

A closed thermofluid system consists of a cylinder and piston assembly and contains 0.5 kg of nitrogen. The piston is loaded so that the gas undergoes a constant pressure process at 250 kPa. The volume changes from 0.2 to 0.5 m³ during the process. Find the heat transferred during the process assuming that $C_P = 1.04 \, \text{kJ/kg K}$.

The constant pressure process can be portrayed on a process diagram, as shown in Figure 12.4.

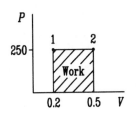

**Fig. 12.4**

*Solution*

From equation (12.9):

$$q = C_P(T_2 - T_1)$$

The values of $T_1$ and $T_2$ can be found using the equation of state, (12.4):

$$P = \rho R T$$

and

$$T = \frac{P}{\rho R}$$

From equation (12.6):

$$R = \frac{R_0}{M} = \frac{8.314}{28} = 0.297 \, \text{kJ/kg K}$$

At state 1,

$$\rho_1 = \frac{m_1}{V_1} = \frac{0.5}{0.2} = 2.5 \, \text{kg/m}^3$$

and

$$T_1 = \frac{250 \times 10^3}{0.297 \times 10^3 \times 2.5} = 336.7 \, \text{K}$$

At state 2,

$$\rho_2 = \frac{m_2}{V_2} = \frac{0.5}{0.5} = 1 \, \text{kg/m}^3$$

and

$$T_2 = \frac{250 \times 10^3}{0.297 \times 10^3 \times 1} = 841.8 \, \text{K}$$

Therefore

$$q = C_P(T_2 - T_1)$$
$$= 1.04(841.8 - 336.7) = 525.3 \, \text{kJ/kg}$$

However, in this example, the mass of gas is 0.5 kg, so the actual heat transferred is

$$525.3 \times 0.5 = 262.6 \, \text{kJ}$$

### 12.4.3 Relationship between the specific heats

The definition of enthalpy, from section 11.6.3, is

$$h = u + \frac{P}{\rho}$$

For a small change of enthalpy, there will be a corresponding change in the internal energy and the work done:

$$\Delta h = \Delta u + \Delta\left(\frac{P}{\rho}\right) \qquad (12.11)$$

From the equation of state (12.4) the work done can be expressed as

$$\Delta\left(\frac{P}{\rho}\right) = R\Delta T$$

From the definitions of $C_V$ and $C_P$, given in equations (12.8) and (12.10), for an ideal gas:

$$\Delta u = C_V\Delta T$$
$$\Delta h = C_P\Delta T$$

Substituting in equation (12.11):

$$C_P\Delta T = C_V\Delta T + R\Delta T$$

Cancelling out the change of temperature, $\Delta T$, gives a relationship between the specific heats for an ideal gas:

$$C_P = C_V + R \qquad (12.12)$$

**Example 12.7**

When 1 kg of an ideal gas is raised from 30 to 80 °C at constant pressure, the heat input is 100 kJ. When the same gas is raised through the same temperature range at constant volume, the heat input is 80 kJ. Find the gas constant for the ideal gas.

*Solution*

For the constant pressure process, using equation (12.9):

$$C_P = \frac{q}{T_2 - T_1} = \frac{100}{80 - 30} = 2\,\text{kJ/kg K}$$

For the constant volume process, using equation (12.7):

$$C_V = \frac{q}{T_2 - T_1} = \frac{80}{80 - 30} = 1.6\,\text{kJ/kg K}$$

From equation (12.12):

$$R = C_P - C_V$$
$$= 2 - 1.6 = 0.4\,\text{kJ/kg K}$$

## 12.5  NON-FLOW PROCESSES FOR AN IDEAL GAS

There are several processes that an ideal gas can undergo within a closed thermofluid system, all of which can be portrayed on a $P-V$ diagram. Four such processes are shown in Figure 12.5 and are labelled a to

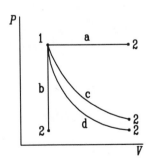

**Fig. 12.5** Non-flow processes.

d respectively. The processes shown are:

> a = constant pressure process
> b = constant volume process
> c = isothermal process
> d = adiabatic process

Constant volume and constant pressure processes have already been discussed in sections 12.4.1 and 12.4.2 respectively. Isothermal and adiabatic processes are discussed below.

### 12.5.1  Isothermal process for an ideal gas

An isothermal process is one that takes place at constant temperature. Therefore, for an ideal gas, the process obeys Boyle's law as defined in equation (12.1):

$$PV = \text{constant}$$

In practice, a perfect isothermal process is difficult to achieve because the work output, or input, for the process has to be matched by an equal amount of heat being transferred across the boundary. This can be seen by considering the non-flow energy equation for an isothermal process. From equation (11.6)

$$q - w = u_2 - u_1$$

However, from equation (12.8)

$$u_2 - u_1 = C_V(T_2 - T_1)$$

Since the temperature is constant,

$$T_2 - T_1 = 0$$

it follows that

$$u_2 - u_1 = 0$$

Hence, for an isothermal process the work done equals the heat transfer since

$$q - \omega = 0$$

The nearest approach to an isothermal process is where the process takes place in a cylinder and piston assembly, and the piston is slow moving, so that heat transfer can take place between the ideal gas in the closed thermofluid system and the surroundings.

### 12.5.2  Adiabatic process for an ideal gas

An adiabatic process is one in which there is no heat transfer between the working fluid in the closed thermofluid system and the surroundings. In the case of the working fluid being an ideal gas, the adiabatic process obeys the relationship

$$PV^\gamma = \text{constant} \tag{12.13}$$

where $\gamma$ is the adiabatic index for expansion or compression.

The index $\gamma$ (gamma) is discussed in more detail in the next section. For the moment, it is necessary to appreciate that within an adiabatic process not only do the pressure and volume change, but the temperature of the gas also changes. The relationship between the changes in pressure and temperature can be derived in the following manner.

From the equation of state (12.4)

$$P = \rho RT$$

so that

$$\rho = \frac{P}{RT}$$

For 1 kg of an ideal gas

$$\rho = \frac{1}{V}$$

and

$$V = \frac{RT}{P} \tag{12.14}$$

The adiabatic relationship can be expressed as

$$P_1 V_1^\gamma = P_2 V_2^\gamma$$

and substituting for $V$ from equation (12.14)

$$P_1 \left( \frac{RT_1}{P_1} \right)^\gamma = P_2 \left( \frac{RT_2}{P_2} \right)^\gamma$$

Cancelling out the gas constant, $R$, and rearranging the pressure and temperature terms

$$\frac{P_1 T_1^\gamma}{P_1^\gamma} = \frac{P_2 T_2^\gamma}{P_2^\gamma}$$

which becomes

$$\frac{T_1^\gamma}{P_1^{\gamma-1}} = \frac{T_2^\gamma}{P_2^{\gamma-1}}$$

Therefore

$$\left( \frac{P_1}{P_2} \right)^{\gamma-1} = \left( \frac{T_1}{T_2} \right)^\gamma$$

or, more conveniently,

$$\frac{P_1}{P_2} = \left( \frac{T_1}{T_2} \right)^{\gamma/(\gamma-1)} \tag{12.15}$$

In practice, an adiabatic process can be nearly achieved if the process is very rapid and there is no time for heat transfer to take place. An adiabatic process approximates closely to the expansion and compression processes in a petrol engine in a car. Similarly an adiabatic process can take place if the closed system is insulated from the surroundings.

### Example 12.8

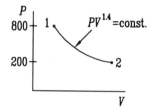

Fig. 12.6

An ideal gas at a pressure of 800 kPa and temperature of 500 K expands adiabatically to a pressure of 200 kPa (Figure 12.6). Calculate the work done for each 1 kg of gas during the process if $\gamma = 1.4$ and $C_V = 0.718 \, \text{kJ/kg K}$.

*Solution*

From the non-flow energy equation (11.6)

$$q - w = u_2 - u_1$$

but $q = 0$, so that

$$w = u_1 - u_2$$

From equation (12.8)

$$w = C_V(T_1 - T_2)$$

In order to find $T_2$ it is necessary to use equation (12.15):

$$\frac{P_1}{P_2} = \left(\frac{T_1}{T_2}\right)^{\gamma/(\gamma-1)}$$

This can be rearranged to the form

$$\frac{T_1}{T_2} = \left(\frac{P_1}{P_2}\right)^{(\gamma-1)/\gamma}$$

$$= \left(\frac{800}{200}\right)^{0.4/1.4} = 1.486$$

Therefore

$$T_2 = \frac{T_1}{1.486} = \frac{500}{1.486} = 336.5 \, \text{K}$$

The work done is

$$w = C_V(T_1 - T_2)$$

$$= 0.718 \times 10^2 (500 - 336.5)$$

$$= 117.4 \times 10^2 \, \text{J/kg} = 117.4 \, \text{kJ/kg}$$

*Note:* the work done is positive, indicating a work output from the expansion process.

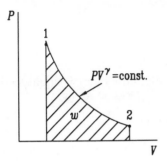

**Fig. 12.7** Adiabatic process.

### 12.5.3 Adiabatic index

It was shown in section 11.6.2 that the work done during a frictionless expansion, or compression, process is equal to the area under the process curve on a $P-V$ diagram. The area under an adiabatic process curve for an ideal gas is illustrated in Figure 12.7. By integration it can be shown that the area under the curve $PV^\gamma = $ constant is

$$w = \frac{P_2 V_2 - P_1 V_1}{1 - \gamma} \tag{12.16}$$

For 1 kg of gas, this can be changed to

$$w = \frac{RT_2 - RT_1}{1 - \gamma}$$

$$= \frac{R(T_1 - T_2)}{\gamma - 1} \tag{12.17}$$

However, in Example 12.8 it was shown that for an adiabatic process

$$w = C_V(T_1 - T_2) \tag{12.18}$$

For both equations (12.17) and (12.18) to be correct it is necessary that

$$C_V = \frac{R}{\gamma - 1}$$

which gives

$$\gamma = \frac{R}{C_V} + 1$$

$$= \frac{R + C_V}{C_V}$$

But, from equation (12.12)

$$R + C_V = C_P$$

so that

$$\gamma = \frac{C_P}{C_V} \tag{12.19}$$

In other words, the adiabatic index is equal to the ratio of the specific heats, and this is true for all ideal gases.

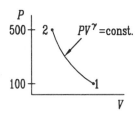

**Fig. 12.8**

### Example 12.9

Hydrogen ($H_2$) is compressed from a pressure of 100 kPa and a temperature of 15 °C to a pressure of 500 kPa (Figure 12.8). Find the temperature of the compressed gas, assuming the process to be adiabatic. Take $C_P = 14.3$ kJ/kg K for hydrogen.

*Solution*

The adiabatic index can be found using equation (12.19):

$$\gamma = \frac{C_P}{C_V}$$

From equation (12.12)

$$C_V = C_P - R$$

and from equation (12.6)

$$R = \frac{R_0}{M}$$

The molecular weight of $H_2$ is $M = 2$, so that

$$R = \frac{8.314}{2} = 4.157 \text{ kJ/kg K}$$

Therefore

$$C_V = 14.3 - 4.157 = 10.143 \text{ kJ/kg K}$$

and

$$\gamma = \frac{14.3}{10.143} = 1.41$$

The temperature of the compressed gas can be found using equation (12.15):

$$\frac{T_2}{T_1} = \left(\frac{P_2}{P_1}\right)^{(\gamma - 1)/\gamma}$$

$$= \left(\frac{500}{100}\right)^{0.41/1.41} = 1.597$$

Therefore

$$T_2 = 1.597 \times (15 + 273) = 459.9 \text{ K}$$

## 12.6 LIQUIDS AND VAPOURS

Liquids and vapours are considered together because they are inextricably linked. A typical example is water. Under normal atmospheric conditions, it exists as a liquid. Raising the temperature to its boiling point causes the water to change to steam, a vapour. When all the water has been converted to steam, additional heat input causes the temperature to rise above the boiling point into what is called the 'superheat' region.

These different states of a fluid are called 'phases'. During boiling, or condensing, a fluid such as water can exist in a state in which both liquid and vapour are present. Under these conditions the fluid is referred to as being in a 'two-phase' state.

### 12.6.1  A fluid undergoing a phase change

Consider a cylinder and frictionless piston assembly, as shown in Figure 12.9(a), containing 1 kg of water at room temperature. The cylinder and piston form a closed thermofluid system. With the piston loaded as shown, the pressure inside the system is greater than atmospheric and will be constant throughout the process.

As heat is transferred to the water, the temperature rises until it reaches boiling point, state 1 on the curve shown in Figure 12.9(b). Thereafter, continuing heat transfer causes boiling to take place at constant temperature until, at state 2, the water is completely changed to steam. From this condition, continuing heat input to the system causes the temperature of the steam to increase in the superheat region to state 3.

The curve shown in Figure 12.9(b) is typical of any fluid going through a phase change. Take any ideal gas and, if the temperature is reduced low enough until it liquefies, it will exhibit exactly the same behaviour as shown in Figure 12.9(b).

Although Figure 12.9(b) provides a useful illustration of what happens to a fluid changing from a liquid to a vapour, or from a vapour to a liquid, it is not a property diagram of the process. This is because the heat input is not a property of the fluid but a measure of the quantity of energy transferred across the system boundary.

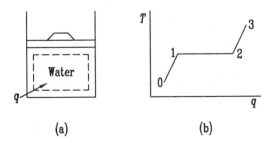

**Fig. 12.9** Water undergoing a phase change.

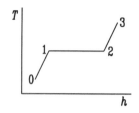

**Fig. 12.10** Phase change drawn on a *T–h* diagram.

The process illustrated in Figure 12.9(b) was at constant pressure. From the discussion in section 11.6.3 it was shown that the heat transfer to, or from, a closed thermofluid system can be equated to the change of enthalpy of the fluid. This is defined in equation (11.8):

$$q = h_2 - h_1$$

where the enthalpy is related to the internal energy, pressure and density of the fluid:

$$h = u + \frac{P}{\rho}$$

It follows that a change of phase can be represented on a temperature–enthalpy diagram, *T–h*, as shown in Figure 12.10.

### 12.6.2 Two-phase fluid conditions

At the beginning of the nineteenth century, one of the theories regarding heat was that it was a fluid that pervaded an object or substance. When heat entered, the object or substance became 'saturated' with this colourless, odourless, weightless fluid. Even though heat has since been recognized as being a form of energy, the expression 'saturated' still remains and is used to describe the condition of a fluid undergoing a two-phase process.

At a particular pressure, a liquid will change to a vapour at one constant temperature. Although this temperature has been referred to as the boiling point, in engineering thermodynamics it is called the **saturation temperature**. The saturation temperature is defined as the temperature at which a change of phase can take place. Clearly, for a given fluid, the saturation temperature occurs at a particular pressure. For example, water at standard atmospheric pressure has a saturation temperature of 100 °C. Either the pressure or the saturation temperature can be used, as a property, to define a fluid undergoing a change of phase.

When a liquid reaches the saturation temperature, it has reached the maximum temperature at which it can exist as a liquid. At this point, it is called a 'saturated liquid'. Similarly, if all the saturated liquid is converted to vapour, then the vapour at saturation temperature is called the 'saturated vapour', as shown in Figure 12.11.

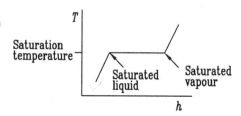

**Fig. 12.11** Saturation conditions.

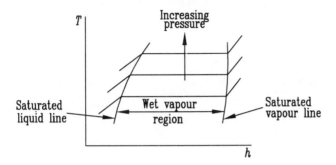

**Fig. 12.12** Wet vapour region.

Between the saturated liquid and saturated vapour states, the fluid will be at saturated temperature but consist of a proportion of liquid and a proportion of vapour in equilibrium. This condition is termed a 'wet vapour'. The steam flowing out of the spout of a kettle is a wet vapour, otherwise it could not be seen. It is the droplets of saturated water in the steam that make it visible.

In order for a saturated liquid to be changed to a saturated vapour, energy must enter the system in the form of heat. The amount of energy required depends on the saturation temperature of the fluid and this, in turn, depends on the pressure of the fluid. As the pressure increases, so does the saturation temperature. It is possible to draw a series of two-phase changes on the $T$–$h$ diagram, as shown in Figure 12.12.

### 12.6.3 Dryness fraction

The state of a fluid can be defined by two independent properties. In the case of a wet vapour, both the saturation temperature and the pressure are dependent on each other, so it is only necessary to use one of these as an independent property. One further independent property is required to define the state of a wet vapour. This is termed the 'dryness fraction' and is given the symbol $x$. (Some textbooks also refer to the dryness fraction as the quality of the vapour.)

The dryness fraction is the ratio of saturated vapour in a mixture of liquid and vapour, based on mass:

$$x = \frac{\text{mass of saturated vapour}}{\text{mass of wet vapour}} \tag{12.20}$$

It follows that, for every kilogram of wet vapour, the mass of saturated vapour is $x$ kilograms and the mass of saturated liquid is $(1 - x)$ kg. Hence, it follows that saturated vapour has a dryness fraction of 1 and saturated liquid has a dryness fraction of 0.

### Example 12.10

One kilogram of wet vapour has a dryness fraction of 0.8 (Figure 12.13). If the enthalpy of the saturated liquid is 418 kJ/kg and the enthalpy of the saturated vapour is 2675 kJ/kg, find the enthalpy of the wet vapour.

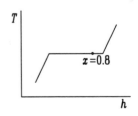

**Fig. 12.13**

*Solution*

The respective masses of saturated liquid and saturated vapour in the wet vapour can be found using equation (12.20):

$$x = \frac{\text{mass of saturated vapour}}{\text{mass of wet vapour}}$$

For 1 kg of wet vapour, $x = 0.8$, so that

mass of saturated vapour $= 0.8$ kg

It follows that:

$$\text{mass of saturated liquid} = 1 - 0.8 = 0.2 \text{ kg}$$
$$\text{enthalpy of saturated vapour} = 0.8 \times 2675 = 2140 \text{ kJ}$$
$$\text{enthalpy of saturated liquid} = 0.2 \times 418 = 83.6 \text{ kJ}$$
$$\text{enthalpy of the wet vapour} = 2140 + 83.6$$
$$= 2223.6 \text{ kJ/kg}$$

**Example 12.11**

A wet vapour has a dryness fraction of 0.6. The density of the saturated liquid is 1000 kg/m³ and the density of the saturated vapour is 1.65 kg/m³. Find the density of the wet vapour.

*Solution*

By definition:

$$\rho = \frac{m}{V}$$

so that the density can be found by finding the volume occupied by 1 kg of wet vapour:

$$\text{mass of saturated vapour} = 0.6 \text{ kg}$$
$$\text{mass of saturated liquid} = 0.4 \text{ kg}$$
$$\text{volume of saturated vapour} = 0.6 \times \frac{1}{1.65} = 0.364 \text{ m}^3$$
$$\text{volume of saturated liquid} = 0.4 \times \frac{1}{1000} = 0.0004 \text{ m}^3$$
$$\text{density of the wet liquid} = \frac{0.6 + 0.4}{0.364 + 0.0004} = 2.744 \text{kg/m}^3$$

*Note:* the volume occupied by the saturated liquid is very much smaller than that of the saturated vapour. Therefore, in calculating the density of a wet vapour the contribution of the saturated liquid can generally be ignored. The actual density can be estimated from:

density of saturated vapour/dryness fraction

### 12.6.4 Superheated vapour conditions

Once a fluid has reached the saturated vapour condition, any further heat input will cause the temperature of the vapour to rise into the **superheat** region. Figure 12.11 shows the variation of temperature in the superheat region for a fluid under constant pressure conditions.

The temperature of a superheated vapour is independent of the pressure. Therefore, pressure and temperature can be used to define the state of a superheated vapour, as they represent two independent properties of the fluid.

## 12.7  USE OF VAPOUR TABLES

The properties of wet, saturated and superheated vapours can be found from published tables, such as in Howatson, Lund and Todd (1991) and Rogers and Mayhew (1988). Since the properties of steam are most widely used in analysing thermofluid situations, these tables tend to be colloquially referred to as 'steam tables'.

For convenience, a shortened version of the property tables for saturated water, saturated steam and superheated steam is given in Appendices C and D. The properties quoted are the water/steam pressure $P$, saturation temperature $T_s$, density $\rho$, internal energy $u$ and enthalpy $h$.

When considering the properties of a wet vapour, it is cumbersome to continually refer to either saturated vapour or saturated liquid. The accepted notation is to define the properties of a saturated liquid by the subscript f and those of a saturated vapour by the subscript g. Consequently, the internal energy and enthalpy for a saturated liquid are

$$u_f, \quad h_f$$

and those for a saturated vapour are

$$u_g, \quad h_g$$

Using the dryness fraction $x$, together with the saturated properties, the properties of a wet vapour can be defined. The dryness fraction can be defined in terms of the mass of the saturated liquid $m_f$ and the mass of the saturated vapour $m_g$:

$$x = \frac{m_g}{m_f + m_g} \tag{12.21}$$

For 1 kg of wet vapour, the mass of the saturated vapour becomes

$$m_g = x$$

and the mass of the saturated liquid

$$m_f = 1 - x$$

As explained in Example 12.11, the volume of a liquid is negligible compared with that of a saturated vapour. Therefore, the density of a wet vapour can be estimated from

$$\rho = \frac{\rho_g}{x} \qquad (12.22)$$

The internal energy of a wet vapour can be found from the sum of the internal energy of the saturated liquid together with the internal energy of the saturated vapour:

$$u = (1 - x)u_f + x \times u_g$$

Rearranging gives

$$u = u_f + x(u_g - u_f) \qquad (12.23)$$

(It should be noted that values of $u_g$ and $u_f$ are quoted in Appendix C. Some textbooks and tables of properties give the difference $(u_g - u_f)$ as $u_{fg}$, with appropriate values.)

Similarly, the enthalpy of a wet vapour is given by the relationship

$$h = h_f + x(h_g - h_f) \qquad (12.24)$$

### Example 12.12

In an insulated piston and cylinder assembly, steam expands from a pressure of 2 MPa and temperature of 250 °C down to a pressure of 200 kPa and dryness fraction of 0.9 (Figure 12.14). Calculate the work done during this process for each kilogram of steam.

*Solution*

The work done can be found using the non-flow energy equation (11.6):

$$q - w = u_2 - u_1$$

The system is insulated so that

$$q = 0$$

Therefore

$$w = u_1 - u_2$$

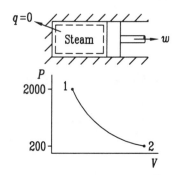

**Fig. 12.14**

From Appendix D, at 2 MPa and 250 °C the steam is superheated (as indicated by the temperature) and

$$u_1 = 2680 \, \text{kJ/kg}$$

From Appendix C, at 200 kPa:

$$u_\text{f} = 505 \, \text{kJ/kg}$$
$$u_\text{g} = 2529 \, \text{kJ/kg}$$

From equation (12.23)

$$u_2 = u_\text{f} + x_2 \, (u_\text{g} - u_\text{f})$$
$$= 505 + 0.9 \, (2529 - 505) = 2326.6 \, \text{kJ/kg}$$

Substituting in the non-flow energy equation:

$$w = 2680 - 2326.6 = 353.4 \, \text{kJ/kg}$$

## 12.8 SUMMARY

Key equations that have been introduced in this chapter are as follows.
The equation of state for an ideal gas:

$$P = \rho R T \tag{12.4}$$

where

$$R = \frac{R_0}{M} \tag{12.6}$$

For an ideal gas undergoing a constant volume process

$$q = C_V(T_2 - T_1) \tag{12.7}$$

where

$$C_V = \frac{\Delta u}{\Delta T} \tag{12.8}$$

For an ideal gas undergoing a constant pressure process:

$$q = C_P(T_2 - T_1) \tag{12.9}$$

where

$$C_P = \frac{\Delta h}{\Delta T} \tag{12.10}$$

Relationship between specific heats for an ideal gas:

$$C_P = C_V + R \tag{12.12}$$

For an ideal gas undergoing an adiabatic process:

$$PV^\gamma = \text{constant} \tag{12.13}$$

$$\frac{P_1}{P_2} = \left(\frac{T_1}{T_2}\right)^{\gamma/(\gamma-1)} \tag{12.15}$$

where

$$\gamma = \frac{C_P}{C_V} \tag{12.19}$$

For a wet vapour:

$$x = \frac{m_g}{m_f + m_g} \tag{12.21}$$

$$\rho = \frac{\rho_g}{x} \tag{12.22}$$

$$u = u_f + x(u_g - u_f) \tag{12.23}$$

$$h = h_f + x(h_g - h_f) \tag{12.24}$$

## 12.9 PROBLEMS

### 12.9.1 Ideal gases

1. A closed vessel contains methane, $M = 16$, at a pressure of 500 kPa and a temperature of 15 °C. If the vessel has a fixed volume of $0.5\,\text{m}^3$, find the mass of gas contained.

2. A rigid container holds 1 kg of an ideal gas at a temperature of 20 °C and pressure of 150 kPa. If the gas has a $C_V$ of 0.75 kJ/kg K, calculate the heat transfer necessary to increase the temperature to 80 °C and the final pressure of the gas.

3. An electric heater of 2 kW is placed inside a closed room $3 \times 4 \times 2\,\text{m}$. If the air is initially at a temperature of 10 °C and pressure of 101 kPa, find the time needed for the air temperature to reach 20 °C. Assume there is no heat loss from the room. Take the properties of air as:

   $R = 0.287\,\text{kJ/kg K}$
   $C_V = 0.72\,\text{kJ/kg K}$

4. Air at a pressure of 100 kPa and temperature of 15 °C is compressed adiabatically to a pressure of 800 kPa. Calculate the work done for each kilogram of air if $\gamma = 1.4$ and $C_V = 0.72\,\text{kJ/kg K}$.

5. Carbon monoxide (CO) is expanded adiabatically from a temperature of 500 K and pressure of 1 MPa, down to a temperature of 250 K. Find the final pressure of the gas and the work done for each kilogram of gas during the process. Take $C_P = 1.04\,\text{kJ/kg K}$ for carbon monoxide.

## 12.9.2 Vapour properties

6. For steam at a pressure of 1 MPa find:

   (a) the internal energy if $x = 0.7$;
   (b) the dryness fraction if $h = 2100\,\mathrm{kJ/kg\,K}$.

7. Within a closed thermofluid system steam is raised from a dryness fraction of 0.3 to a temperature of 250 °C at a constant pressure of 2 MPa. Calculate the change of enthalpy of the steam.

8. Within a cylinder and piston assembly a quantity of steam is raised from a dryness fraction of 0.4 to 0.85 at a constant temperature of 86 °C. Find the heat transfer to the system for each kilogram of steam.

9. Calculate the work done during the process described in problem 8 for each kilogram of steam.

10. At a pressure of 200 kPa 1 kg of wet steam has a volume of 0.1 m³. Find the dryness fraction and the internal energy of the steam.

# Steady flow of fluids 13

## 13.1 AIMS

- To introduce the continuity equation for steady flow through an open thermofluid system.
- To introduce the momentum equation for steady flow.
- To introduce the steady flow energy equation and discuss its application to: boilers and condensers; turbines and compressors; nozzles.
- To introduce Bernoulli's equation for ideal incompressible flow.
- To analyse velocity-measuring devices using Bernoulli's equation.

## 13.2 OPEN THERMOFLUID SYSTEM

An open thermofluid system has already been introduced in section 1.4.2 as a region into, and out of, which there is a flow of fluid across the boundary defining the system. The basic features of an open thermofluid system are shown in Figure 13.1. It consists of a region in space that is defined by means of the boundary. Across the boundary there will be a flow of fluid and, in addition, there can be transfer of energy in the form of heat and work.

The same sign convention is used as for the closed thermofluid system, described in section 11.6, with: heat input to an open system being positive; work output from an open system being positive.

Just as for the closed system, the boundary of an open thermofluid system plays an essential part in defining the system. It provides a boundary between the fluid within this system and the surroundings.

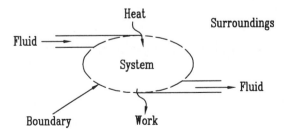

**Fig. 13.1** Open thermofluid system.

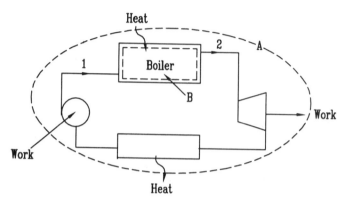

**Fig. 13.2** Steam plant.

This can sometimes take the form of a physical boundary, as in the case of a steam boiler, where the boiler shell forms a solid boundary between the hot water/steam inside and the surrounding atmospheric air. The boundary also provides a necessary visual model for problem solving by defining the geometry of the particular region under consideration. By defining the boundary, the open thermofluid system is defined.

This can be visualized by considering the steam plant shown in Figure 13.2. It is similar to the steam plant shown in Figure 11.23 and consists of a boiler, turbine, condenser and feed pump. By drawing the boundary around the complete plant, boundary A defines a closed thermofluid system because the contained fluid is a fixed quantity. However, by drawing the boundary around just one of the components as, for example, boundary B which envelops the boiler, the new system is open because there is flow of fluid across the boundary at points 1 and 2.

The open thermofluid systems considered in this book operate with 'steady' flow. This means that the flow out of the system equals the flow in. In other words, there is no build-up or reduction of fluid within the actual system. Also, the flow rate does not vary with time. In order to analyse open thermofluid systems, it is necessary to apply one, or more, of the following equations:

1. continuity
2. momentum
3. energy.

These are discussed in the following sections.

### 13.3 CONTINUITY EQUATION

Continuity in this context is defined as a state of continuous and steady flow. For steady flow across an open thermofluid system the mass flow

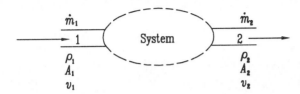

**Fig. 13.3** Steady flow across an open system.

rate of fluid entering the system must equal the mass flow rate leaving the system.

Consider the open system shown in Figure 13.3. For steady flow the mass flow rate at 1 equals the mass flow rate at 2:

$$\dot{m}_1 = \dot{m}_2$$

Now the fluid at 1 has a particular density, $\rho_1$, and enters with velocity $v_1$ through a flow area $A_1$. The volume flow rate is velocity multiplied by the flow area $(A_1 \times v_1)$. This can be seen by considering the units:

$$m^2 \times \frac{m}{s} = \frac{m^3}{s}$$

i.e. volume/time. The volume flow rate can be converted to a mass flow rate by multiplying by the density $\rho_1$:

$$\dot{m} = \rho_1 \times A_1 \times v_1$$

and, again, this can be seen by considering the units:

$$\frac{kg}{m^3} \times m^2 \times \frac{m}{s} = \frac{kg}{s}$$

i.e. mass/time. A similar argument can be used to show that the mass flow rate at the exit is

$$\dot{m} = \rho_2 \times A_2 \times v_2$$

Continuity is, therefore, achieved when

$$\dot{m} = \rho_1 A_1 v_1 = \rho_2 A_2 v_2 \qquad (13.1)$$

### Example 13.1

In a water fountain, the water leaves a vertical nozzle of 25 mm diameter with a velocity of 10 m/s. Assuming the water flow to remain circular, find the diameter of the flow at a height of 2.5 m above the nozzle. Take the density of water as 1000 kg/m³ and $g = 9.81$ m/s².

The fountain can be pictured as an open system as shown in Figure 13.4.

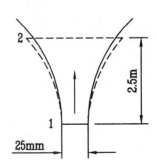

**Fig. 13.4**

*Solution*

The water flow can be considered to consist of a large number of droplets, each of which must obey the laws of motion. The velocity at 2 can be found using equation (2.5):

$$v_2^2 = v_1^2 + 2as$$
$$= v_1^2 - 2gs$$
$$= (10)^2 - 2 \times 9.81 \times 2.5$$

Therefore $v_2 = 7.14 \, \text{m/s}$.

From continuity equation (13.1)

$$\rho_1 A_1 v_1 = \rho_2 A_2 v_2$$

but since the density of water is constant:

$$A_1 v_1 = A_2 v_2$$

$$\frac{\pi}{4}(0.025)^2 \times 10 = \frac{\pi}{4}(d_2)^2 \times 7.14$$

Therefore $d_2 = 29.6 \, \text{mm}$.

## 13.4 MOMENTUM EQUATION

A steady flow of fluid can exert a force on its surroundings and this can be analysed by applying Newton's second law. This can be expressed as the **rate of change of momentum** of the flow. As explained in section 1.5.1, momentum is the product of mass × velocity, so Newton's second law can be written in the form of equation (1.2):

$$F = \frac{\Delta(mv)}{\Delta t}$$

where $F$ is the force resulting from the rate of change of momentum.

For a steady flow of fluid the change of mass with respect to time, $\Delta m / \Delta t$, can be expressed as the mass flow rate, $\dot{m}$. Equation (1.2) can be rewritten in the form

$$F = \dot{m}v \qquad (13.2)$$

which is the 'momentum' equation for steady fluid flow.

One application of the momentum is in propulsive devices such as propellers, for both waterborne craft and aircraft. It also applies to 'jet' engines. For both types of device, a propulsive force is created if the fluid leaving the device is at a greater velocity than the velocity entering, as shown in Figure 13.5. Figure 13.5(a) shows a propeller with fluid entering at velocity $v_1$ and leaving at velocity $v_2$. Figure 13.5(b) shows a similar situation for a jet engine.

In both cases there will be propulsive force created as shown provided that

$$v_2 > v_1$$

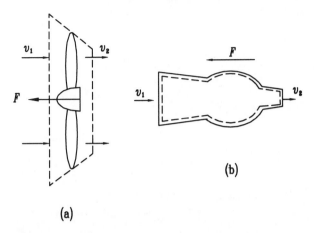

**Fig. 13.5** Propulsive devices.

At the entry to either device there will be a resultant force with a magnitude of $\dot{m}v_1$ acting towards the right. At the exit there will be a resultant force with a magnitude of $\dot{m}v_2$ acting towards the left. Since

$$v_2 > v_1$$

it follows that

$$\dot{m}v_2 > \dot{m}v_1$$

and there is a resultant force acting towards the left with a magnitude

$$F = \dot{m}v_2 - \dot{m}v_1$$
$$= \dot{m}(v_2 - v_1) \tag{13.3}$$

**Example 13.2**

A boat travelling at 6 m/s in sea water has a 250 mm diameter propeller that discharges the water at a velocity of 12 m/s. Assuming the effect of the propeller hub to be negligible, find the magnitude of the thrust produced. Assume the density of sea water to be 1030 kg/m³. (See Figure 13.5(a).)

*Solution*

The mean water velocity through the propeller is

$$\bar{v} = \frac{6 + 12}{2} = 9 \, \text{m/s}$$

From continuity equation (13.1)

$$\dot{m} = \rho A \bar{v}$$

$$= 1030 \times \frac{\pi}{4}(0.25)^2 \times 9 = 455 \, \text{kg/s}$$

The thrust can be found using equation (13.3):

$$F = \dot{m}(v_2 - v_1)$$
$$= 455\,(12 - 6) = 2730\,\text{N}$$

## 13.5 ENERGY EQUATION

Energy must be conserved, so any energy transfer across the boundary of an open thermofluid system must be equal to the change of energy between the outlet and inlet of the system. This can be expressed as

transfer of energy to system = energy leaving system
− energy entering system

It is possible to evaluate the energy entering and leaving the system by considering the open thermofluid system shown in Figure 13.6. The fluid enters at section 1 with its own particular properties such as internal energy, pressure and density. However, these three properties can be combined into one property 'enthalpy'. Thus, the energy content of the fluid entering the system comprises: enthalpy, kinetic energy and potential energy.

For each kilogram of fluid entering at section 1 the total energy is

$$h_1 + \tfrac{1}{2}v_1^2 + z_1 g$$

Similarly, the total energy of each kilogram of fluid at section 2 is

$$h_2 + \tfrac{1}{2}v_2^2 + z_2 g$$

The difference between these energy levels is given by the transfer of energy across the boundary of the system in the form of heat $q$ and work $w$. The energy equation for an open system can be expressed as

$$q - w = \left(h_2 + \frac{v_2^2}{2} + z_2 g\right) - \left(h_1 + \frac{v_1^2}{2} + z_1 g\right) \tag{13.4}$$

Because this equation is based on steady flow conditions, it is called the steady flow energy equation, SFEE for short.

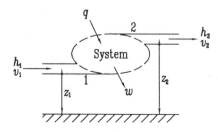

**Fig. 13.6** Energy conservation for an open system.

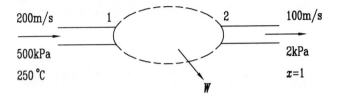

**Fig. 13.7**

## Example 13.3

A steady flow device (Figure 13.7) is insulated and operates with steam entering at a pressure of 500 kPa, a temperature of 250 °C and a velocity of 200 m/s. If the steam leaves the device as saturated vapour at a pressure of 200 kPa, with a velocity of 100 m/s, find the work done for each kilogram of steam. Ignore any change of height across the device.

*Solution*

The work done can be found using the steady flow energy equation (13.4):

$$q - w = \left( h_2 + \frac{v_2^2}{2} + z_2 g \right) - \left( h_1 + \frac{v_1^2}{2} + z_1 g \right)$$

This can be simplified because

$q = 0$   (insulated)

$z_1 = z_2$

Therefore

$$w = \left( h_1 + \frac{v_1^2}{2} \right) - \left( h_2 + \frac{v_2^2}{2} \right)$$

From Appendix D

$h_1 = 2961 \, \text{kJ/kg} = 2961 \times 10^3 \, \text{J/kg}$

From Appendix C

$h_2 = 2706 \, \text{kJ/kg} = 2706 \times 10^3 \, \text{J/kg}$

Therefore

$$w = 2961 \times 10^3 + \frac{(200)^2}{2} - 2706 \times 10^3 - \frac{(100)^2}{2}$$

$$= 270\,000 \, \text{J/kg} = 270 \, \text{kJ/kg}$$

*Note:* each term of the SFEE must be in consistent units. Taking the kinetic energy term, $v^2/2$, the units are: $\text{m}^2/\text{s}^2$. Multiplying both top and bottom by "kg" gives

$$\frac{\text{kg m}^2}{\text{kg s}^2} = \frac{\text{kg m}}{\text{s}^2} \frac{\text{m}}{\text{kg}} = \frac{\text{N m}}{\text{kg}} = \frac{\text{J}}{\text{kg}}$$

It follows that all other terms in the equation must be expressed in the same units.

## 13.6 STEADY FLOW THERMOFLUID DEVICES

There is a wide range of steady flow devices used in mechanical engineering. Boilers, condensers, turbines, compressors and nozzles are but some of them. In order to limit the discussion, it is proposed to illustrate the application of the steady flow energy equation to just these few devices in the following sections.

### 13.6.1 Boilers and condensers

A boiler is a device that brings about a change of phase due to a heat input. In the case of a steam boiler, water enters the shell and is converted into steam. A condenser is the reverse of the boiler; a vapour enters the condenser and, due to heat transfer away from the fluid, is condensed to a liquid.

A schematic diagram of a boiler is shown in Figure 13.8, but this could equally well describe a condenser except that in the case of a condenser there is heat transfer from the fluid.

The energy changes across a boiler, or condenser, and can be analysed using the steady flow energy equation (13.4):

$$q - w = \left( h_2 + \frac{v_2^2}{2} + z_2 g \right) - \left( h_1 + \frac{v_1^2}{2} + z_1 g \right)$$

This can be simplified by making the following assumptions:

1. There is no work done, so that $w = 0$.
2. The change of velocity across the device is small, so that $v_1^2 \simeq v_2^2$.
3. The change of height across the device is small, so that $z_1 \simeq z_2$.

Using these assumptions, the steady flow energy equation becomes

$$q = h_2 - h_1 \qquad (13.5)$$

and this equation is valid for both a boiler and a condenser.

### Example 13.4

A boiler operates at a pressure of 1 MPa. Water enters in a saturated condition and leaves with a dryness fraction of 0.8. Calculate the heat transfer to the boiler for each kilogram of water converted to wet steam. (See Figure 13.8.)

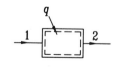

**Fig. 13.8** Diagram of a boiler.

*Solution*

The heat transfer can be found using equation (13.5):

$$q = h_2 - h_1$$

At state 2, from Appendix C

$$h_f = 763\,\text{kJ/kg} \qquad h_g = 2776\,\text{kJ/kg}$$

From equation (12.24)

$$h_2 = h_f + x_2(h_g - h_f)$$
$$= 763 + 0.8(2776 - 763) = 2373.4\,\text{kJ/kg}$$

At state 1, from Appendix C

$$h_1 = 763\,\text{kJ/kg}$$

Therefore

$$q = (2373.4 - 763) \times 10^3$$
$$= 1610.4 \times 10^3\,\text{J/kg} = 1610.4\,\text{kJ/kg}$$

### 13.6.2 Turbines and compressors

Turbines are devices that produce a positive work output due to expansion of a fluid from a high pressure to a low pressure. The type of turbine is defined by the fluid flowing through the system and can generally be categorized as either a steam or a gas turbine.

Compressors, by comparison, can be considered as the reverse of turbines in that the purpose is to raise the pressure of a fluid – generally a gas – by means of a negative work input.

Figure 13.9 shows the schematic diagrams representing both a turbine and a compressor. It will be seen that one is the mirror image of the other and, in principle, this is a useful way to view the two devices. Assuming there is no heat loss to the surroundings and the flow processes are without friction, then the turbine can, in principle, be reversed to work as a compressor and the compressor reversed to work as a turbine. This allows the analysis of the turbine and compressor using the steady flow energy equation (13.4):

$$q - w = \left( h_2 + \frac{v_2^2}{2} + z_2 g \right) - \left( h_1 + \frac{v_1^2}{2} + z_1 g \right)$$

This can be simplified by making the following assumptions:

1. There is no heat transfer across the boundary, so that $q = 0$.
2. The change of velocity across the device is small, so that $v_1^2 \simeq v_2^2$.
3. The change of height across the device is small, so that $z_1 \simeq z_2$.

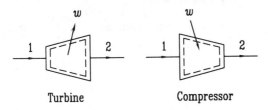

Turbine        Compressor

**Fig. 13.9** Diagrams of a turbine and a compressor.

Using these assumptions, the steady flow energy equation becomes

$$-w = h_2 - h_1$$

which can be rearranged into the form

$$w = h_1 - h_2 \qquad\qquad (13.6)$$

and this equation is valid for both a turbine and a compressor.

It should be noted that when either a turbine or a compressor operates with an ideal gas, the enthalpy values can be found from equation (12.10):

$$h = C_p(T)$$

The use of this equation is illustrated in Example 13.6 below.

### Example 13.5

Steam enters a turbine at a pressure of 2 MPa and a temperature of 300 °C. If it expands down to a pressure of 50 kPa and dryness fraction of 0.88, find the work done for each kilogram of steam flowing through the turbine. (See Figure 13.9.)

*Solution*

The work done can be found using equation (13.6):

$$w = h_1 - h_2$$

At state 1, from Appendix D

$$h_1 = 3025 \text{ kJ/kg}$$

At state 2, from Appendix C

$$h_f = 341 \text{ kJ/kg} \qquad h_g = 2646 \text{ kJ/kg}$$

From equation (12.24)

$$h_2 = h_f + x_2(k_g - h_f)$$
$$= 341 + 0.88(2646 - 341) = 2369.4 \text{ kJ/kg}$$

Therefore

$$w = (3025 - 2369.4) \times 10^3$$
$$= 655.6 \times 10^3 \text{ J/kg} = 655.6 \text{ kJ/kg}$$

*Note:* the work done is positive, indicating a work output from the turbine.

### Example 13.6

Air enters a compressor at a temperature of 15 °C and pressure of 100 kPa. If it is compressed to a pressure of 700 kPa, find the work done for each kilogram of air. For air, take $\gamma = 1.4$ and $C_P = 1.005\,\text{kJ/kg K}$. (See Figure 13.9.)

*Solution*

The work done can be found using equation (13.6):

$$w = h_1 - h_2$$

Since air can be considered to be an ideal gas, from equation (12.10)

$$w = C_P(T_1 - T_2)$$

Since there is no heat transfer, the process can be assumed to be adiabatic, with the relationship between temperature and pressure being given by equation (12.15):

$$\frac{P_1}{P_2} = \left(\frac{T_1}{T_2}\right)^{\gamma/(\gamma - 1)}$$

Rearranging

$$T_2 = T_1\left(\frac{P_2}{P_1}\right)^{(\gamma - 1)/\gamma}$$

Now $T_1 = 15 + 273 = 288\,\text{K}$, therefore

$$T_2 = 288\left(\frac{700}{100}\right)^{0.4/1.4} = 502.2\,\text{K}$$

The work done is

$$w = 1.005 \times 10^3(288 - 502.2)$$
$$= -215.2 \times 10^3\,\text{J/kg} = -215.2\,\text{kJ/kg}$$

*Note:* the negative sign for the work means that it is a work input to the compressor.

### 13.6.3 Nozzles

Nozzles are devices for increasing the velocity of a flow by expanding the fluid from a high pressure down to a lower pressure. In steam turbines, nozzles are used to accelerate the flow of steam before it enters the rotating turbine blades. In jet engines, nozzles are used to accelerate the flow of hot gas to produce the rate of change of momentum necessary to create the propulsive thrust.

Figure 13.10 shows a schematic diagram representing a nozzle. The flow through a nozzle can be analysed using the steady flow energy equation (13.4):

$$q - w = \left(h_2 + \frac{v_2^2}{2} + z_2 g\right) - \left(h_1 + \frac{v_1^2}{2} + z_1 g\right)$$

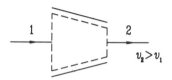

**Fig. 13.10** Diagram of a nozzle.

This can be simplified by making the following assumptions

1. There is no heat transfer across the boundary, so that $q = 0$.
2. There is no work done by the nozzle, so that $w = 0$.
3. The change of height across the nozzle is small, so that $z_1 \simeq z_2$.

   In addition, it is generally accepted that the inlet area of a nozzle is large, so there is a further assumption that:

4. The inlet velocity to the nozzle is low, so that $v_1 \simeq 0$.

Using these assumptions, the steady flow energy equation becomes

$$0 = h_2 + \frac{v_2^2}{2} - h_1$$

which can be rearranged into the form

$$v_2 = \sqrt{[2(h_1 - h_2)]} \tag{13.7}$$

It should be noted that, when a nozzle operates with an ideal gas, the enthalpy values can be found from equation (12.10):

$$h = C_P(T)$$

### Example 13.7

In a jet engine hot gas enters a nozzle at a pressure of 400 kPa and a temprature of 800 K. If the gas expands down to a pressure of 100 kPa, calculate the velocity at the outlet of the nozzle. For the gas, take $\gamma = 1.38$ and $C_P = 1.05$ kJ/kg K. (See Figure 13.10.)

*Solution*

The outlet velocity can be found using equation (13.7):

$$v_2 = \sqrt{[2(h_1 - h_2)]}$$

Since the fluid flowing through the nozzle is a gas, this relationship can be modified using equation (12.10):

$$v_2 = \sqrt{[2C_P(T_1 - T_2)]}$$

Since there is no heat transfer from the nozzle, the flow process can be assumed to be adiabatic with the relationship between temperature and pressure being given by equation (12.15):

$$\frac{P_1}{P_2} = \left(\frac{T_1}{T_2}\right)^{\gamma/(\gamma-1)}$$

Rearranging

$$T_2 = T_1\left(\frac{P_2}{P_1}\right)^{(\gamma-1)/\gamma}$$

$$= 800\left(\frac{100}{400}\right)^{0.38/1.38} = 546.1 \text{ K}$$

The outlet velocity is

$$v_2 = \sqrt{[2 \times 1.05 \times 10^3(800 - 546.1)]}$$
$$= 730.2 \text{ m/s}$$

*Note:* it is important to convert the value of $C_P$ to units of J/kg K in order for equation (13.7) to be valid.

## 13.7 THE STEADY FLOW ENERGY EQUATION AS A RATE EQUATION

The steady flow energy equation, as expressed in equation (13.4), applies to just 1 kg of fluid flowing through the system. There is nothing to show whether the 1 kg flows through the system in 1 s or 1 h. Equation (13.4) cannot be used to find the rate of change of energy for the system. Fortunately, it can be modified quite easily to provide a rate equation by introducing the mass flow rate.

Taking the steady flow energy equation (13.4):

$$q - w = \left(h_2 + \frac{v_2^2}{2} + z_2g\right) - \left(h_1 + \frac{v_1^2}{2} + z_1g\right)$$

and multiplying throughout by the mass flow rate $\dot{m}$ gives

$$\dot{m}q - \dot{m}w = \dot{m}\left(h_2 + \frac{v_2^2}{2} + z_2g\right) - \dot{m}\left(h_1 + \frac{v_1^2}{2} + z_1g\right)$$

However, the rate of heat transfer, $\dot{m}q$, can be expressed as $Q$, and the rate of work done, $\dot{m}w$, can be expressed as $W$. As a consequence, the rate equation becomes

$$Q - W = \dot{m}\left(h_2 + \frac{v_2^2}{2} + z_2q\right) - \dot{m}\left(h_1 + \frac{v_1^2}{2} + z_1g\right) \tag{13.8}$$

Since the terms in the steady flow energy equation are expressed in units of J/kg, the terms in the rate equation (13.8) are expressed in units of

$$\frac{\text{kg}}{\text{s}} \times \frac{\text{J}}{\text{kg}} = \frac{\text{J}}{\text{s}} = \text{W}$$

i.e. watts.

The rate equation (13.8) can be used to find the rate of heat transfer in boilers or condensers. Modifying equation (13.5) gives the rate of heat transfer as

$$Q = \dot{m}(h_2 - h_1) \tag{13.9}$$

Similarly, the rate equation can be used to find the rate of work done for turbines or compressors. Now the rate of work done is power and the power output of a turbine, or the power input to a compressor, can be found using a modified form of equation (13.6):

$$W = \dot{m}(h_1 - h_2) \tag{13.10}$$

The application of the rate equation (13.8) is further illustrated in Example 13.8 below, which also emphasizes the usefulness of this form of the steady flow energy equation.

### Example 13.8

In a hydroelectric system, water flows from a reservoir through an inclined circular pipe down to a water turbine situated 250 m below the level of the reservoir (Figure 13.11). If the pipe is 0.5 m in diameter and the water flows at a velocity of 2 m/s, find the power produced by the system. Take $g = 9.81 \, \text{m/s}^2$ and the density of water, $\rho = 1000 \, \text{kg/m}^3$.

*Solution*

The power can be found using equation (13.8):

$$Q - W = \dot{m}\left(h_2 + \frac{v_2^2}{2} + z_2 g\right) - \dot{m}\left(h_1 + \frac{v_1^2}{2} + z_1 g\right)$$

This can be simplified by making the following assumptions:

1. no heat transfer, $Q = 0$;
2. no change in water temperature, $h_1 = h_2$;
3. no change of velocity, $v_1 = v_2$.

The rate equation becomes

$$-W = \dot{m}z_2 g - \dot{m}z_1 g$$

Rearranging

$$W = \dot{m}(z_1 g - z_2 g)$$

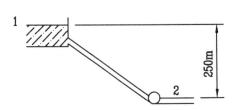

**Fig. 13.11**

The mass flow rate can be found using equation (13.1):

$$\dot{m} = \rho A v$$

$$= 1000 \times \frac{\pi}{4}(0.5)^2 \times 2 = 392.7 \, \text{kg/s}$$

The power output from the system is

$$W = 392.7(250 \times 9.81 - 0)$$
$$= 963\,096.8 \, \text{W} = 963.1 \, \text{kW}$$

*Note:* in other words, the power output equals the rate of change of potential energy for the system.

## 13.8  BERNOULLI'S EQUATION

Bernoulli's equation represents the form of the steady flow energy equation applied to ideal, incompressible flow with no heat transfer to, or from, the fluid and no work done on, or by, the fluid. As such, it is a useful equation for studying a range of fluid flow situations, as outlined in the following discussion.

Taking the steady flow energy equation (13.4):

$$q - w = \left( h_2 + \frac{v_2^2}{2} + z_2 g \right) - \left( h_1 + \frac{v_1^2}{2} + z_1 g \right)$$

this can be simplified by making the following assumptions:

1. There is no heat transfer across the boundary, so that $q = 0$.
2. There is no work done by the fluid, so that $w = 0$.

As a result, the steady flow energy equation can be written as

$$0 = \left( h_2 + \frac{v_2^2}{2} + z_2 g \right) - \left( h_1 + \frac{v_1^2}{2} + z_1 g \right)$$

or

$$h_1 + \frac{v_1^2}{2} + z_1 g = h_2 + \frac{v_2^2}{2} + z_2 g \tag{13.11}$$

Now the definition of enthalpy is

$$h = u + \frac{P}{\rho}$$

which, when substituted in equation (13.11), gives

$$u_1 + \frac{P_1}{\rho_1} + \frac{v_1^2}{2} + z_1 g = u_2 + \frac{P_2}{\rho_2} + \frac{v_2^2}{2} + z_2 g \tag{13.12}$$

However, ideal flow means flow without friction. Since there is no friction and both $q$ and $w$ are zero, there is no change in temperature of the fluid and assumption 3 can be stated as:

3. The flow is frictionless, so that $u_2 = u_1$.

Finally, the fluid is taken to be incompressible, in other words the density of the fluid does not change:

4. The fluid is incompressible, so that

$$\rho_1 = \rho_2 = \rho \quad \text{(a constant)}$$

Using these assumptions, equation (13.12) becomes

$$\frac{P_1}{\rho} + \frac{v_1^2}{2} + z_1 g = \frac{P_2}{\rho} + \frac{v_2^2}{2} + z_2 g \tag{13.13}$$

in which form it is **Bernoulli's equation**.

Bernoulli's equation can be applied to flow of any fluid, providing that the fluid remains incompressible. In the case of ideal gases, this is approximately true for flow velocities of less than 100 m/s.

The form of Bernoulli's equation given in equation (13.13) is an energy equation in which each term represents an energy content of the flow with units of J/kg. It is clear that

$$\frac{v^2}{2} = \text{kinetic energy}$$

$$zg = \text{potential energy}$$

However, it is not quite so clear what form of energy is represented by the $P/\rho$ term. That it is an energy is evident from the units:

$$\frac{\text{Pa}}{\text{kg/m}^3} = \frac{\text{N m}^3}{\text{m}^2 \text{ kg}} = \frac{\text{N m}}{\text{kg}} = \frac{\text{J}}{\text{kg}}$$

In reality, the term is defined as

$$\frac{P}{\rho} = \text{flow work}$$

What this means in practice is that fluid entering an open thermofluid system has work done upon it by the fluid pushing behind. Similarly, the fluid leaving has to do work by pushing the fluid ahead of it. This work is the **flow work**.

As well as an energy equation, Bernoulli's equation can be expressed as a pressure equation. By multiplying equation (13.13) throughout by the density, $\rho$, the resulting equation is

$$P_1 + \rho\frac{v_1^2}{2} + \rho z_1 g = P_2 + \rho\frac{v_2^2}{2} + \rho z_2 g \tag{13.14}$$

Each of these terms is a pressure, defined as

$$P = \text{static pressure}$$

$$\rho\frac{v^2}{2} = \text{dynamic pressure}$$

$$\rho z g = \text{hydrostatic pressure}$$

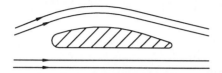

**Fig. 13.12** Flow around an aerofoil.

A typical example of the application of equation (13.14) is shown in Figure 13.12, flow around an aerofoil. An aerofoil is the cross-section of an aircraft wing. In practice, a wing moves through the air in order to generate 'lift'. It is easier to visualize the flow around an aerofoil by considering the aerofoil to be stationary and the air to be flowing around it.

Continuity requires that the air flowing over the top of the aerofoil section has a higher velocity than that flowing underneath, because the air has to travel a greater distance from the leading edge to the trailing edge. Due to the higher velocity, the static pressure of the air over the top of the aerofoil must be less than that underneath. This vertical difference in pressure across the aerofoil, when acting over the whole wing, results in an upward force which is the 'lift'.

### Example 13.9

A light aircraft has a wing area of 15 m² and flies at 60 m/s in level flight. If the air velocity over the lower surface of the wing is equal to the flight speed and the velocity over the upper surface is 15% greater, find the magnitude of the lift developed. Take the density of air as 1.2 kg/m³. (See Figure 13.12.)

*Solution*

Applying Bernoulli's equation (13.14) to flow around the wing, it can be assumed that there is negligible change in height across the aerofoil, so that

$$P_\mathrm{u} + \rho \frac{v_\mathrm{u}^2}{2} = P_1 + \rho \frac{v_1^2}{2}$$

where subscript u refers to the upper surface and l the lower surface of the wing. The pressure difference is

$$P_1 - P_\mathrm{u} = \frac{\rho}{2}(v_\mathrm{u}^2 - v_1^2)$$

$$= \frac{1.2}{2}((60 \times 1.15)^2 - 60^2) = 696.6 \, \mathrm{Pa}$$

The lift is the pressure difference multiplied by the wing area:

$$L = 696.6 \times 15 = 10\,449 \, \mathrm{N}$$

### 13.9 FLOW MEASUREMENT

There are several devices for measuring the velocity of fluid flows that operate with a pressure difference and are capable of being analysed using Bernoulli's equation. These include the pitot tube, venturi-meter and orifice plate as described below.

#### 13.9.1 Pitot tube

In its simplest form, the pitot tube is an open-ended tube pointing into the direction of the flow and connected to some form of pressure gauge or manometer. A pitot tube of this type is shown in Figure 13.13. The velocity of the free stream at section 1 is $v_1$. At the entry to the pitot tube, the velocity at section 2 will be zero because there is no flow through the tube. Applying equation (13.14) to this situation:

$$P_1 + \rho \frac{v_1^2}{2} + \rho z_1 g = P_2 + \rho \frac{v_2^2}{2} + \rho z_2 g$$

Assuming there is no change of height, so that $z_1 = z_2$, and $v_2 = 0$, the equation becomes

$$P_1 + \rho \frac{v_1^2}{2} = P_2$$

Rearranging gives

$$\rho \frac{v_1^2}{2} = P_2 - P_1$$

and

$$v_1 = \sqrt{\left[ \frac{2(P_2 - P_1)}{\rho} \right]} \qquad (13.15)$$

#### Example 13.10

In a wind tunnel the air velocity is measured by means of a pitot tube. If the pressure difference is recorded with a manometer reading 12 mm height of water, calculate the air velocity. Take the water density as 1000 kg/m³, air density as 1.2 kg/m³ and $g = 9.81$ m/s². (See Figure 13.13.)

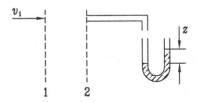

**Fig. 13.13** Pitot tube.

*Solution*

The pressure difference can be evaluated using the hydrostatic equation (11.2):

$$P_2 - P_1 = \rho z g$$
$$= 1000 \times 0.012 \times 9.81 = 117.7 \, \text{Pa}$$

Substituting this value in equation (13.15):

$$v_1 = \sqrt{\left(\frac{2(P_2 - P_1)}{\rho}\right)} = \sqrt{\frac{2 \times 117.7}{1.2}} = 14 \, \text{m/s}$$

### 13.9.2 Venturi-meter

A venturi-meter is a device that can be incorporated in a pipe to measure the flow velocity and, hence, the mass flow rate. Its basic form is shown in Figure 13.14 and involves a reduction in flow area to create a measurable pressure difference $(P_1 - P_2)$.

If the cross-sectional flow areas at sections 1 and 2 are $A_1$ and $A_2$ respectively, then the velocities at these two sections can be related through the continuity equation (13.1):

$$\rho_1 A_1 v_1 = \rho_2 A_2 v_2$$

For incompressible fluids $\rho$ is constant, so that

$$\rho_1 = \rho_2$$

and

$$v_2 = \frac{A_1}{A_2} v_1$$

Applying equation (13.14) to the venturi-meter:

$$P_1 + \rho \frac{v_1^2}{2} + \rho z_1 g = P_2 + \rho \frac{v_2^2}{2} + \rho z_2 g$$

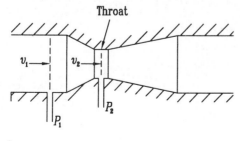

**Fig. 13.14** Venturi-meter.

Assuming there is no change of height, so that $z_1 = z_2$, the equation becomes

$$P_1 + \rho\frac{v_1^2}{2} = P_2 + \rho\frac{v_2^2}{2}$$

Substituting for $v_2$

$$P_1 + \rho\frac{v_1^2}{2} = P_2 + \rho\frac{((A_1/A_2)v_1)^2}{2}$$

which can be rearranged to give

$$v_1 = \sqrt{\left[\frac{2(P_1 - P_2)}{\rho((A_1/A_2)^2 - 1)}\right]} \tag{13.16}$$

In practice, this gives a reasonably accurate value of flow velocity, within 5% of the true value, for a well-designed venturi-meter.

### Example 13.11

A venturi-meter is used to measure the flow of water through a pipe. The diameters of the pipe and throat are 100 and 50 mm respectively. If the measured pressure difference between the inlet and throat is 150 mm of water, estimate the mass flow rate through the pipe. Take the density of water as 1000 kg/m³ and $g = 9.81$ m/s². (See Figure 13.14.)

*Solution*

The pressure difference can be evaluated using the hydrostatic equation (11.2):

$$P_1 - P_2 = \rho z g$$
$$= 1000 \times 0.15 \times 9.81 = 1471.5 \, \text{Pa}$$

The ratio of the areas is

$$\frac{A_1}{A_2} = \frac{\pi/4(d_1^2)}{\pi/4(d_2^2)} = \left(\frac{d_1}{d_2}\right)^2 = \left(\frac{100}{50}\right)^2 = 4$$

Substituting in equation (13.16)

$$v_1 = \sqrt{\left[\frac{2(P_1 - P_2)}{\rho((A_1/A_2)^2 - 1)}\right]}$$

$$= \sqrt{\frac{2 \times 1471.5}{1000(4^2 - 1)}} = 0.443 \, \text{m/s}$$

The mass flow rate can be found from the continuity equation (13.1):

$$\dot{m} = \rho_1 A_1 v_1 = 1000 \times \frac{\pi}{4}(0.1)^2 \times 0.443$$

$$= 3.48 \, \text{kg/s}$$

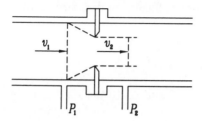

**Fig. 13.15** Ideal flow through an orifice plate.

### 13.9.3 Orifice plate

An orifice plate is another device that allows the velocity of flow in a pipe to be found, by placing a restriction in the flow as shown in Figure 13.15. An orifice plate is convenient to install in a pipeline because, as implied in Figure 13.15, such a plate can be installed at a flanged joint.

The restriction causes a measurable pressure drop $(P_1 - P_2)$. For ideal flow leaving the orifice plate, the flow cross-section area will be the same as for the orifice, $A_2$, with the same velocity of $v_2$. This core of moving fluid is assumed to be surrounded by an annular region of stationary fluid. At section 2 both the moving and stationary fluid are considered to have the same static pressure, $P_2$.

On this basis, the analysis of the orifice plate is the same as for the venturi-meter, resulting in the same equation (13.16) as the relationship for the velocity at section 1:

$$v_1 = \sqrt{\left[\frac{2(P_1 - P_2)}{\rho((A_1/A_2)^2 - 1)}\right]}$$

However, the flow for an orifice plate does not follow the ideal behaviour shown in Figure 13.15. In the first place, the flow downstream of the orifice does not remain parallel but undergoes a further contraction, as shown in Figure 13.16. Secondly, the flow surrounding the central moving core does not remain stationary but develops flow eddies due to the entrainment of the fluid by the movement of the core.

What this means in practice is that equation (13.16) gives a value of velocity that is noticeably greater than the actual value. It is possible to overcome this by using a 'correction factor' in connection with equation (13.16). This is termed the 'coefficient of discharge' and given the symbol

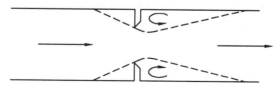

**Fig. 13.16** Actual flow through an orifice plate.

$C_d$. Incorporating the coefficient of discharge into equation (13.16) gives a relationship for the actual velocity at section 1:

$$v_1 = C_d \sqrt{\left[ \frac{2(P_1 - P_2)}{\rho((A_1/A_2)^2 - 1)} \right]} \tag{13.17}$$

In both Figures 13.15 and 13.16 the type of orifice plate illustrated is referred to as 'sharp edged'. Such orifices have well-defined behaviour with a $C_d$ value of around 0.60–0.62.

**Example 13.12**

A sharp-edged orifice plate is installed in a pipe to measure the flow of a gas. The diameters of the pipe and orifice are 150 and 100 mm respectively and the coefficient of discharge is 0.6. If the pressure difference either side of the orifice plate is equivalent to 600 mm of water, find the velocity of the gas through the pipe.

Take the density of water as 1000 kg/m³, the density of the gas as 1.8 kg/m³ and $g = 9.81$ m/s². (See Figure 13.15.)

*Solution*

The pressure difference can be evaluated using the hydrostatic equation (11.2):

$$P_1 - P_2 = \rho z g = 1000 \times 0.6 \times 9.81 = 5886 \text{ Pa}$$

The ratio of the areas is

$$\frac{A_1}{A_2} = \frac{(\pi/4)(d_1)^2}{(\pi/4)(d_2)^2} = \left( \frac{d_1}{d_2} \right)^2 = \left( \frac{150}{100} \right)^2 = 2.25$$

Substituting in equation (13.17)

$$v_1 = C_d \sqrt{\left[ \frac{2(P_1 - P_2)}{\rho((A_1/A_2)^2 - 1)} \right]}$$

$$= 0.6 \sqrt{\left[ \frac{2 \times 5886}{1.8(2.25^2 - 1)} \right]} = 24.1 \text{ m/s}$$

13.10  SUMMARY

Key equations that have been introduced in this chapter are as follows.
    Continuity equation for steady flow:

$$\dot{m} = \rho_1 A_1 v_1 = \rho_2 A_2 v_2 \tag{13.1}$$

Force due to a rate of change of flow momentum:

$$F = \dot{m}v \tag{13.2}$$

Steady flow energy equation:

$$q - w = \left( h_2 + \frac{v_2^2}{2} + z_2 g \right) - \left( h_1 + \frac{v_1^2}{2} + z_1 g \right)$$ (13.4)

Steady flow through a boiler or condenser:

$$q = h_2 - h_1$$ (13.5)

Steady flow through a turbine or compressor:

$$w = h_1 - h_2$$ (13.6)

Exit velocity for a nozzle:

$$v_2 = \sqrt{[2(h_1 - h_2)]}$$ (13.7)

SFEE as a rate equation:

$$Q - W = \dot{m} \left( h_2 + \frac{v_2^2}{2} + z_2 g \right) - \dot{m} \left( h_1 + \frac{v_1^2}{2} + z_1 g \right)$$ (13.8)

Rate equation for a boiler or condenser:

$$Q = \dot{m}(h_2 - h_1)$$ (13.9)

Rate equation for a turbine or compressor:

$$W = \dot{m}(h_1 - h_2)$$ (13.10)

Bernoulli's equation for ideal flow:

$$\frac{P_1}{\rho} + \frac{v_1^2}{2} + z_1 g = \frac{P_2}{\rho} + \frac{v_2^2}{2} + z_2 g$$ (13.13)

or as a pressure equation:

$$P_1 + \rho \frac{v_1^2}{2} + \rho z_1 g = P_2 + \rho \frac{v_2^2}{2} + \rho z_2 g$$ (13.14)

Velocity measured using a pitot tube:

$$v = \sqrt{\left[ \frac{2(P_2 - P_1)}{\rho} \right]}$$ (13.15)

Velocity measured using a venturi-meter:

$$v = \sqrt{\left[ \frac{2(P_1 - P_2)}{\rho((A_1/A_2)^2 - 1)} \right]}$$ (13.16)

Velocity measured using an orifice plate:

$$v = C_d \sqrt{\left[ \frac{2(P_1 - P_2)}{\rho((A_1/A_2)^2 - 1)} \right]}$$ (13.17)

### 13.11 PROBLEMS

1. A compressor takes in air with a density of $1.2 \, \text{kg/m}^3$ at a velocity of $20 \, \text{m/s}$ through a circular duct of $100 \, \text{mm}$ diameter. It discharges the compressed air at a velocity of $12 \, \text{m/s}$ through a duct $50 \, \text{mm}$ square. Find the density of the air at the outlet.

2. A light aircraft has a propeller of $2 \, \text{m}$ diameter and flies at a forward velocity of $50 \, \text{m/s}$. If the exhaust velocity of the propeller is $70 \, \text{m/s}$, calculate the thrust. Take the density of air as $1.2 \, \text{kg/m}^3$.

3. The nozzle of a fireman's hose has an outlet diameter of $50 \, \text{mm}$. It is required to send a jet of water through a vertical height of $25 \, \text{m}$. Calculate the velocity leaving the nozzle and the force on the nozzle. Take the density of water as $1000 \, \text{kg/m}^3$ and $g = 9.81 \, \text{m/s}^2$.

4. A boiler operates at a pressure of $600 \, \text{kPa}$. Water enters at a temperature of $60.1 \, °\text{C}$ and leaves with a dryness fraction of $0.9$. Find the heat transfer to the boiler for each kilogram of water converted to wet steam.

5. A compressor takes in air at $20 \, °\text{C}$ and operates with a pressure ratio of 8. Calculate the work done for each kilogram of air. For air, take $\gamma = 1.4$ and $C_P = 1.005 \, \text{kJ/kg K}$.

6. Steam enters a nozzle with negligible velocity and a temperature of $250 \, °\text{C}$ at a pressure of $500 \, \text{kPa}$. If the steam leaves the nozzle with a pressure of $40 \, \text{kPa}$ and a dryness fraction of $0.94$, find the exit velocity of the steam.

7. Air enters a hair dryer at $20 \, °\text{C}$ and leaves at $60 \, °\text{C}$ through a circular outlet of $50 \, \text{mm}$ diameter. If the outlet velocity is $15 \, \text{m/s}$, calculate the rate of heat input to the dryer. For air, take $\rho = 1.2 \, \text{kg/m}^3$ and $C_P = 1.005 \, \text{kJ/kg K}$.

8. The horizontal carburettor of a petrol engine incorporates a venturi with an inlet diameter of $30 \, \text{mm}$ and a throat diameter of $24 \, \text{mm}$. At the inlet, the air velocity is $25 \, \text{m/s}$ and the static pressure is $100 \, \text{kPa}$. Calculate the static pressure at the throat. Take the density of air as $1.2 \, \text{kg/m}^3$.

9. In a wind tunnel the air velocity is measured using a pitot tube. If the pressure difference recorded with a water manometer is $14 \, \text{mm}$ in height and the air density in the tunnel is $1.25 \, \text{kg/m}^3$, calculate the air velocity. Take the density of water as $1000 \, \text{kg/m}^3$ and $g = 9.81 \, \text{m/s}^2$.

10. An orifice plate is installed in a pipe to measure the flow of water. The diameters of the pipe and orifice are 100 and $50 \, \text{mm}$ respectively and the coefficient of discharge is $0.62$. If the measured pressure difference across the orifice plate is $15 \, \text{mm}$ of mercury, calculate the mass flow rate through the pipe. Take the density of mercury as $13\,600 \, \text{kg/m}^3$, the density of water as $1000 \, \text{kg/m}^3$ and $g = 9.81 \, \text{m/s}^2$.

# Flow with friction $\boxed{14}$

## 14.1 AIMS

- To define viscosity as a property of a fluid.
- To analyse the torque required to overcome friction in a journal bearing.
- To describe the behaviour of flow with friction and the use of the Reynolds number as a criterion for judging the nature of the flow.
- To define the skin friction coefficient for flow with friction.
- To analyse the pressure drop associated with friction in pipes.

## 14.2 LIMITATIONS OF FRICTIONLESS FLOW

Chapter 13 described the behaviour of the frictionless flow of incompressible fluids. As a result, the energy equation for steady flow was defined by Bernoulli's equation (13.13):

$$\frac{P_1}{\rho} + \frac{v_1^2}{2} + z_1 g = \frac{P_2}{\rho} + \frac{v_2^2}{2} + z_2 g$$

This can be applied to steady flow of an incompressible fluid in a long horizontal pipe of constant diameter, as shown in Figure 14.1.

Considering the situation shown in Figure 14.1, there is no change of cross-section, so that $v_1 = v_2$. The pipe is horizontal, so that $z_1 = z_2$. It follows that the static pressures are equal, $P_1 = P_2$, and there is no loss of pressure between sections 1 and 2. This means that there is no driving force to move the fluid along the pipe.

Clearly, this is impossible. In practice, a pressure difference is required between sections 1 and 2 to ensure that there is flow in the direction shown. The magnitude of the pressure difference defines not only the direction of the flow but also the velocity of the flow. The pressure difference is generally referred to as a 'pressure drop'.

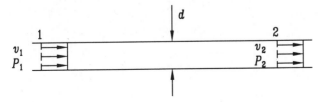

**Fig. 14.1** Frictionless flow in a pipe.

A pressure drop is required to overcome friction in the flow. The friction is created due to the viscous nature of the fluid together with the nature of the flow. These will be discussed later in this chapter. For the present, it can be stated that fluid flow in pipes or ducts can be either laminar or turbulent. In laminar flow the friction is entirely due to the viscous nature of the fluid, defined by the property 'viscosity', which is a measure of the fluid's resistance to flow.

## 14.3 VISCOSITY

One characteristic of the behaviour of real flows with friction is that immediately adjacent to a solid surface the velocity of the flow is zero with respect to that surface. This is because the layer of molecules of the fluid in contact with the surface will adhere to that surface. Moving out from the surface each subsequent layer of fluid will gain increasing velocity as it 'shears' past the lower layers.

Consider the behaviour of a fluid between two parallel solid surfaces, as shown in Figure 14.2. If one surface is stationary, the layer of fluid in contact with it will be stationary. If the other surface is moving with velocity $v$, the fluid in contact with that surface will also have velocity $v$. Across the fluid the velocity will change from 0 to $v$ and, if the gap width $y$ is relatively small, the velocity profile will be as shown in Figure 14.2.

However, to ensure that the moving surface achieves a constant velocity $v$, a force $F$ must be applied such that

$$F \propto A\frac{v}{y}$$

where $A$ is the area in contact with the fluid of the moving surface. This can be re-expressed as

$$\frac{F}{A} \propto \frac{v}{y}$$

But (force/area) is a stress and in this case $(F/A)$ is a shear stress, $\tau$, so that

$$\tau = \text{constant} \left(\frac{v}{y}\right) \tag{14.1}$$

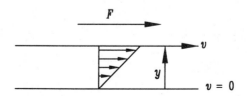

**Fig. 14.2** Fluid between parallel plates.

The constant varies from fluid to fluid, but for any particular fluid at a given temperature the constant can be replaced by the property 'viscosity', given the symbol $\mu$. Equation (14.1) can be restated in terms of the viscosity and the change of velocity in the flow:

$$\tau = \mu\left(\frac{v}{y}\right) \tag{14.2}$$

Fortunately, most fluids used in engineering behave in the manner defined by equation (14.2), in which the shear stress is proportional to the velocity gradient. Such fluids are termed 'Newtonian fluids'. However, there are some fluids which do not behave in the manner defined and are termed 'non-Newtonian fluids'. It is emphasized that only Newtonian fluids will be discussed in this book.

Strictly speaking, the correct title for $\mu$ is the 'dynamic viscosity' but it is referred to as viscosity throughout this book. The units of viscosity can be determined from equation (14.2):

$$\tau = \mu\left(\frac{v}{y}\right)$$

$$\frac{N}{m^2} = \mu\left(\frac{m/s}{m}\right)$$

so the basic dimensions of $\mu$ are

$$\frac{N\,s}{m^2} = \frac{kg\,m}{s^2}\frac{s}{m^2} = kg/m\,s$$

Newtonian fluids have a constant value of viscosity, but $\mu$ is only a constant at a particular temperature. Away from these values, the viscosity will change with temperature. In the case of a liquid the viscosity decreases as the temperature increases. In the case of a gas the viscosity increases with temperature. Typical variations of viscosity with temperature for water and air are given in Table 14.1. It is usually assumed that the viscosity of these two fluids is independent of pressure.

**Table 14.1** Variation of viscosity with temperature

| Temperature (°C) | Water $10^{-3}$ (kg/ms) | Air $10^{-5}$ (kg/ms) |
|---|---|---|
| 0 | 1.75 | 1.71 |
| 20 | 1.00 | 1.81 |
| 40 | 0.65 | 1.90 |
| 60 | 0.46 | 1.99 |
| 80 | 0.35 | 2.08 |
| 100 | 0.28 | 2.17 |

## 14.4 JOURNAL BEARING

It might be thought that the flow situation shown in Figure 14.2 is an artificial one, simply created to be used as a basis for defining viscosity. This is far from the truth; Figure 14.2 models the situation found in a very comon engineering application, the journal bearing in which a shaft runs in a cylindrical housing with a thin layer of lubricant between the two. Since the gap is very small in relation to the size of the shaft, it approximates to fluid between two parallel surfaces.

Figure 14.3 shows a typical journal bearing in which a shaft of diameter $d$ is assumed to rotate concentrically within a layer of fluid, having a constant thickness $y$ between it and the bearing surface. If the shaft rotates at $N$ rev/s, the surface velocity of the shaft will be $(\pi d N)$ and the velocity gradient across the fluid is

$$\frac{v}{y} = \frac{\pi d N}{y}$$

Substituting this in equation (14.2) gives the shear stress at the surface of the shaft as

$$\tau = \mu \left( \frac{\pi d N}{y} \right)$$

The force to create the shear stress is provided by the torque on the shaft:

$$\tau = \frac{T}{(d/2)A} = \frac{T}{(d/2)(\pi d l)}$$

Combining the two equations for the shear stress gives

$$T = \frac{\mu \pi^2 d^3 l N}{2y} \tag{14.3}$$

This can be used to find the power dissipated at the bearing, because the power, $W$, is equal to $(T\omega)$:

$$W = T\omega = T \times 2\pi N \tag{14.4}$$

In practice, there are likely to be radial forces applied to the shaft that would cause a lateral movement in relation to the bearing. Under these circumstances, the film of lubrication would not have a constant thickness, $y$, but would vary around the periphery of the shaft. Nevertheless,

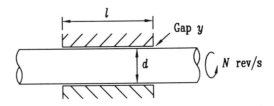

**Fig. 14.3** A journal bearing.

even under these circumstances the torque defined by equation (14.3) can still provide a good approximation to that found in practice.

**Example 14.1**

In an engine the crankshaft rotates at 2000 rev/min in a main journal bearing. If the diameter of the shaft is 30 mm, the width of the bearing is 25 mm and the annular clearance between the shaft and bearing is 0.025 mm, calculate the power dissipated in friction. Assume the viscosity of the lubricating oil to be 0.05 kg/m s. (See Figure 14.3.)

*Solution*

The rotational speed of the shaft,

$$N = \frac{2000}{60} = 33.3 \, \text{rev/s}$$

Applying equation (14.3)

$$T = \frac{\mu \pi^2 d^3 l N}{2y}$$

$$= \frac{0.05 \times \pi^2 \times (0.03)^3 \times 0.025 \times 33.3}{2 \times 0.000\,025}$$

$$= 0.22 \, \text{N m}$$

The power dissipated is given by equation (14.4):

$$W = T \times 2\pi N = 0.22 \times 2\pi \times 33.3 = 46 \, \text{W}$$

## 14.5 FLOW BEHAVIOUR WITH FRICTION

For flow that is not restrained within a narrow gap between two parallel surfaces, the variation of velocity is different from that predicted in Figure 14.2.

The flow of a fluid parallel to a flat surface is shown in Figure 14.4. At the surface the velocity is zero, so that at the start, at section 1, the layer of

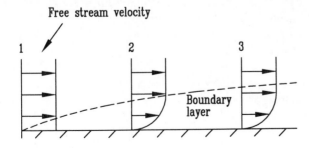

**Fig. 14.4** Development of a boundary layer.

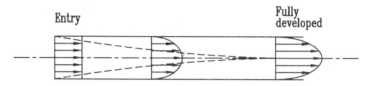

**Fig. 14.5** Developing flow within a pipe.

fluid at the surface is stationary but the remainder of the fluid is uninfluenced by the surface. As the fluid flows along, an increasing number of fluid layers are influenced by the shearing of the fluid, and the resulting velocity profiles are shown at sections 2 and 3. These show that the fluid velocity increases from zero at the surface to the uninfluenced 'free stream' velocity of the flow.

In Figure 14.4 the distance from the surface at which the velocity attains the free stream value is shown by means of a dotted line. The region between the dotted line and the surface is called the 'boundary layer'.

A boundary layer can be defined as the region of flow adjacent to a surface in which the flow velocity varies from zero to the free stream value. In practice, the boundary layer for external flow over a flat plate is fairly thin. However, for flow inside a pipe or duct, the thickness of the boundary layer can be significant, in some cases extending to the centre-line of the flow.

Fluid flow within a pipe falls into one of two separate flow regimes, laminar or turbulent flow. Figure 14.5 shows the development of the boundary layer for laminar flow within a pipe. At the entry to the pipe, the fluid velocity will be uniform. As the flow moves along inside the pipe, the action of viscous shearing will cause the flow to slow down near the pipe wall. However, to maintain continuity, as the flow slows down near the wall, the remainder of the flow near the centre must increase in velocity to above the inlet velocity. When the boundary layer extends from the wall to the centre-line, the flow is said to be 'fully developed' and has a velocity profile that is parabolic. Thereafter, the flow remains fully developed and the parabolic velocity profile is maintained along the remainder of the pipe.

The parabolic profile shown in Figure 14.5 is only true for laminar flow. In the case of turbulent flow the behaviour becomes more inconsistent, with fluid particles moving at random across most of the flow and just a thin boundary layer at the pipe wall. To visualize the difference in behaviour between laminar and turbulent flow, consider the two situations shown in Figure 14.6. This shows the path of a fluid particle for both laminar flow, as shown in Figure 14.6(a), and turbulent flow, as shown in Figure 14.6(b). Within laminar flow, a particle moving along the centre-line of the pipe will continue to move in a steady manner. Within turbulent flow the movement becomes unsteady and then the particle will move in a completely random manner. Although the particle shown in Figure 14.6(b) moves downstream within this general movement, it can

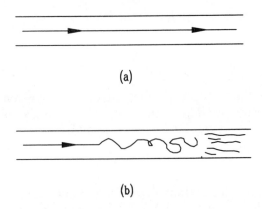

(a)

(b)

**Fig. 14.6** Flow behaviour in a pipe.

move transversely across the flow and can even move upstream. It must be appreciated that this individual particle will be surrounded by other fluid particles, all moving at random.

A typical example of the difference between laminar and turbulent flow can be seen in the smoke rising from a lighted candle. The smoke rises vertically just above the flame, laminar behaviour, then it becomes unsteady and eventually disperses, turbulent behaviour.

What determines whether the flow in a pipe is laminar or turbulent is the mean velocity of the flow, the pipe diameter and the properties of the fluid. These can be combined into a group of variables known as the 'Reynolds number' and given the symbol Re.

## 14.6 REYNOLDS NUMBER

The Reynolds number is an important criterion in the study of fluid flow and is defined in terms of the pipe diameter, $d$, mean flow velocity, $\bar{v}$, and the fluid properties of density, $\rho$, and viscosity, $\mu$:

$$\mathrm{Re} = \frac{\rho \bar{v} d}{\mu} \tag{14.5}$$

In terms of their dimensions, the four variables in this group can be expressed as

$$\frac{(\mathrm{kg/m^3})(\mathrm{m/s})(\mathrm{m})}{\mathrm{kg/m\,s}} = 1$$

Dimensions of '1' in this case means that the group is dimensionless.

Dimensionless groups are a useful means of comparing different flow geometries and fluids. For flow in a pipe, laminar flow can only exist up to a Reynolds number value of 2000. This is true irrespective of the size of pipe or the fluid being considered. Above this value instabilities occur in the laminar flow and transition to turbulent flow takes place. For flow in pipes the various flow regimes can be summarized as:

- laminar flow, $\text{Re} \leqslant 2000$
- transition flow, $2000 < \text{Re} < 4000$
- turbulent flow, $\text{Re} \geqslant 4000$

In practice, the majority of flow situations are within the turbulent regime. Laminar flow only occurs with highly viscous liquids in small pipes.

**Example 14.2**

Water flows through a 200 mm diameter pipe at a mean velocity of 1 m/s. Calculate the Reynolds number for this flow. Find the mean velocity of air inside a 50 mm diameter pipe to achieve the same Reynolds number.
    Take the properties of water and air as:

- water: $\rho = 1000 \, \text{kg/m}^3$, $\mu = 10^{-3} \, \text{kg/m s}$;
- air: $\rho = 1.2 \, \text{kg/m}^3$, $\mu = 1.8 \times 10^{-5} \, \text{kg/m s}$.

A diagram is inappropriate as only the value of the Reynolds number is required to assess whether the flow is laminar or turbulent.

*Solution*

The Reynolds number for flow in a pipe can be found from equation (14.5):

$$\text{Re} = \frac{\rho \bar{v} d}{\mu}$$

For the flow of water:

$$\text{Re} = \frac{1000 \times 1 \times 0.2}{10^{-3}} = 200\,000 \quad \text{i.e. turbulent.}$$

For the flow of air:

$$200\,000 = \frac{1.2 \times \bar{v} \times 0.05}{1.8 \times 10^{-5}}$$

Therefore $\bar{v} = 60 \, \text{m/s}$.

*Note:* at first sight, the water flow and air flow would appear completely different, but the behaviour and velocity profiles are identical due to their having the same Reynolds number.

## 14.7   PRESSURE DROP IN PIPES

The pressure drop for flow in a pipe depends on whether the flow is laminar or turbulent. Since these flow regimes are determined by the Reynolds number of the flow, it is obvious that the pressure drop is also dependent on the Reynolds number. However, it is inconvenient to use the Reynolds number direcly as there is no general relationship between the pressure drop and Reynolds number for all flow regimes. Instead, the Reynolds number is taken to be a function of $f$, which is the 'skin friction coefficient' for the flow.

### 14.7.1 Skin friction coefficient

The skin friction coefficient, as the name implies, is a means of relating the friction between the pipe wall and the fluid flow. It is based on the following relationship for the shear stress at the wall of the pipe:

$$\tau_w = f(\tfrac{1}{2}\rho\bar{v}^2) \tag{14.6}$$

where $\tau_w$ is the shear stress at the pipe wall and $f$ is the skin friction coefficient.

From the discussion in section 13.8, it will be seen that the term $(\tfrac{1}{2}\rho\bar{v}^2)$ represents the **dynamic pressure** of the flow. Therefore, the skin friction coefficient can be defined as the ratio of the shear stress at the wall to the dynamic pressure of the flow.

Since both the shear stress and the dynamic pressure have the same units, Pa, it follows that the skin friction coefficient is **dimensionless.**

Ignoring entry effects, the shear stress at the pipe wall can be considered to be constant along the length of the pipe. Taking the flow of fluid through a pipe of diameter $d$ and length $l$, as shown in Figure 14.7, the forces on the cylindrical element of fluid can be equated as follows:

$$\Delta P\left(\frac{\pi}{4}d^2\right) = \tau_w(\pi dl)$$

Cancelling $(\pi d)$ from both sides of the equation:

$$\Delta P\frac{d}{4} = \tau_w l$$

Substituting for the shear stress, as defined in equation (14.6), gives

$$\Delta P\frac{d}{4} = f\tfrac{1}{2}\rho\bar{v}^2 l$$

which can be rearranged to give a relationship for the pressure drop in the

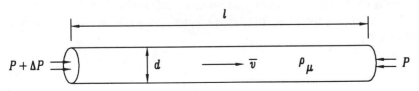

**Fig. 14.7** Shear stress at the pipe wall.

pipe:

$$\Delta P = 2f\frac{l}{d}\rho\bar{v}^2 \tag{14.7}$$

In reality, this equation for the pressure drop in a pipe depends on the definition of the skin friction coefficient. (*Note:* many American textbooks use a skin friction coefficient that is four times greater than that defined above.)

**Example 14.3**

Water flows through a hosepipe of 20 mm diameter and 50 m length, with a mean velocity of 0.5 m/s. Calculate the pressure drop in the pipe due to friction. Take the skin friction coefficient for the pipe as 0.01 and the density of water as 1000 kg/m$^3$. (See Figure 14.7.)

*Solution*

The pressure drop in the pipe can be found using equation (14.7):

$$\Delta P = 2f\frac{l}{d}\rho\bar{v}^2$$

$$= 2 \times 0.01 \times \frac{50}{0.02} \times 1000 \times (0.5)^2$$

$$= 12\,500\,\text{Pa} = 12.5\,\text{kPa}$$

*Note:* the head of water required to overcome this pressure drop can be found from the hydrostatic pressure equation (11.2):

$$\Delta P = \rho g z$$
$$12\,500 = 1000 \times 9.81 \times z$$

and

$$z = 1.27\,\text{m}$$

**14.7.2  Pressure drop in smooth pipes**

Clearly, the pressure drop in a pipe can be calculated once the value of the skin friction coefficient is known. Earlier, it was stated that the pressure drop depends on the Reynolds number of the flow. It follows that the value of the skin friction coefficient is a function of the Reynolds number and depends on whether the flow is laminar or turbulent.

The values of the skin friction coefficient for laminar flow in smooth pipes are given by

$$f = \frac{16}{\text{Re}} \tag{14.8}$$

and for turbulent flow in smooth pipes by

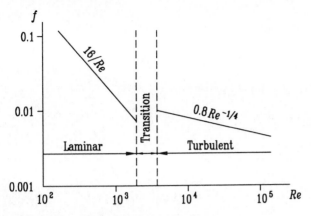

**Fig. 14.8** Skin friction coefficient for smooth pipes.

$$f = 0.08 \, \text{Re}^{-1/4} \tag{14.9}$$

A more detailed discussion of these relationships is given in Sherwin and Horsley (1995).

A plot of skin friction coefficient against Reynolds number is shown in Figure 14.8, using equations (14.8) and (14.9). It can be assumed that laminar flow exists at Reynolds numbers of up to 2000. True turbulent flow does not occur in smooth pipes until the Reynolds number reaches 4000. Within the region

$$2000 < \text{Re} < 4000$$

and the flow will be neither wholly laminar nor wholly turbulent. This is the **transition region** and is the flow regime in which the behaviour changes from laminar to turbulent. If it is necessary to calculate the skin friction coefficient for flow within this region, it is advisable to use equation (14.9) as it is better to err on the side of caution and take a value that is too high rather than one that is too low.

**Example 14.4**

Recalculate the pressure drop for the hosepipe defined in Example 14.3, assuming the pipe to be smooth. Take the properties of water as $\rho = 1000 \, \text{kg/m}^3$ and $\mu = 10^{-3} \, \text{kg/m s}$. (See Figure 14.7.)

*Solution*

The Reynolds number for the flow can be found using equation (14.5):

$$\text{Re} = \rho \frac{\bar{v}d}{\mu}$$

$$= \frac{1000 \times 0.5 \times 0.02}{10^{-3}} = 10\,000$$

The flow is turbulent and the skin friction coefficient can be found from equation (14.9):

$$f = 0.08(\text{Re})^{-1/4}$$
$$= 0.08(10\,000)^{-1/4} = 0.008$$

Applying this value in equation (14.7)

$$\Delta P = 2f\frac{l}{d}\rho\bar{v}^2$$

$$= 2 \times 0.008 \times \frac{50}{0.02} \times 1000 \times (0.5)^2$$

$$= 10\,000\,\text{Pa} = 10\,\text{kPa}$$

### Example 14.5

Oil, having a viscosity of 0.04 kg/m s flows through a 10 mm diameter pipe for a distance of 4 m. If the flow rate is 1 litre/min, find the pressure drop in the pipe due to friction. Assume the flow to be laminar. (See Figure 14.7.)

*Solution*

It will be noticed that the density of the oil is not given. However, it is possible to solve this problem by substituting equation (14.8) for the skin friction coefficient for laminar flow into equation (14.7).
From equation (14.8)

$$f = \frac{16}{\text{Re}} = \frac{16\mu}{\rho\bar{v}d}$$

Substituting in equation (14.7):

$$\Delta P = 2f\frac{l}{d}\rho\bar{v}^2$$

$$= 2\left(\frac{16\mu}{\rho\bar{v}d}\right)\frac{l}{d}\rho\bar{v}^2$$

$$= \frac{32\mu l\bar{v}}{d^2}$$

The volume flow rate is 1 litre/min, therefore

$$\bar{v} = \frac{\text{volume flow rate}}{\text{flow area}}$$

$$= \frac{1/(1000 \times 60)}{\pi/4 \times 0.01^2} = 0.667\,\text{m/s}$$

and

$$\Delta P = \frac{32 \times 0.04 \times 4 \times 0.667}{0.01^2} = 34\,150\,\text{Pa}$$

$$= 34.15\,\text{kPa}$$

### 14.7.3 Pressure drop in rough pipes

In practice, pipes are rarely perfectly smooth and, as a result of their manufacture or construction, have a certain amount of surface roughness that can affect the magnitude of the skin friction coefficient.

It is found that the roughness of a pipe wall has no effect on the skin friction coefficient within laminar flow. However, roughness will have some effect on the value of the skin friction coefficient within turbulent flow. In reality, the roughness causes the skin friction coefficient to be increased above that for a smooth pipe. This is discussed in more detail in Sherwin and Horsley (1995) but, as a 'rule of thumb', a value of

$$f = 0.01$$

is a useful guide to the magnitude of the skin friction coefficient for turbulent flow in rough pipes.

## 14.8 FLOW IN PIPELINE SYSTEMS

To ensure that fluid flows through a pipeline, there must be a power input to overcome the pressure drop due to friction. This power is provided by a pump, in the case of liquids, or a fan, in the case of gases. The combination of the pipeline and pump, or fan, forms a system.

A typical pipeline system is shown in Figure 14.9. There is a power input to the system to overcome not only the pressure drop in the pipe but also the change of height between the entry and exit of the system. It is assumed that the mass flow rate of the fluid through the system is constant.

The power input to the system can be found by using the steady flow energy equation as a rate equation. From equation (13.8)

$$Q - W = \dot{m}\left( h_2 + \frac{v_2^2}{2} + z_2 g \right) - \dot{m}\left( h_1 + \frac{v_1^2}{2} + z_1 g \right)$$

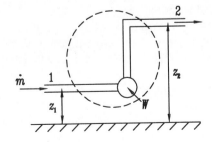

**Fig. 14.9** Pipeline system.

This can be simplified by making the following assumptions:

1. There is no heat transfer across the boundary, so that $Q = 0$.
2. There is no change of velocity across the system, so that $v_1 = v_2$.

Using these assumptions, the steady flow rate equation becomes

$$-W = \dot{m}(h_2 + z_2 g) - \dot{m}(h_1 + z_1 g)$$

which can be rearranged into the form

$$W = \dot{m}((h_1 - h_2) + g(z_1 - z_2)) \tag{14.10}$$

Now the definition of enthalpy is

$$h = u + \frac{P}{\rho}$$

which, when substituted in equation (14.10), gives

$$W = \dot{m}\left((u_1 - u_2) + \left(\frac{P_1}{\rho_1} - \frac{P_2}{\rho_2}\right) + g(z_1 - z_2)\right) \tag{14.11}$$

The change of internal energy $(u_1 - u_2)$ represents the work needed to be done to overcome friction in the pipeline system. It can be re-expressed as an additional flow work term $(-\Delta P/\rho)$, the negative sign indicating that work needs to be done on the system to overcome friction.

For an incompressible fluid the density is constant and equation (14.11) becomes

$$W = \dot{m}\left(\left(\frac{P_1 - P_2}{\rho}\right) - \frac{\Delta P}{\rho} + g(z_1 - z_2)\right)$$

Multiplying throughout by $\rho$ gives

$$W = \frac{\dot{m}}{\rho}(P_1 - P_2 - \Delta P + \rho g(z_1 - z_2))$$

which can be expressed as

$$W = \frac{\dot{m}}{\rho}\Delta P_0 \tag{14.12}$$

where $\Delta P_0$ is the change of **total pressure** of the system and combines the change of static pressure $(P_1 - P_2)$, pressure drop $\Delta P$ and change of hydrostatic pressure $\rho g(z_1 - z_2)$. The application of equation (14.12) is demonstrated in Example 14.6 below.

It should be noted that equation (14.12) is applicable to any pipeline system, including systems in which there is a power output, as in the case of the hydroelectric system considered in Example 14.7 below.

### Example 14.6

A steam boiler produces 1000 kg/h of steam at 500 kPa (Figure 14.10). Water is fed to the boiler from a source at 100 kPa by means of a pump. The water flows through a pipe of 25 mm diameter and 10 m length. Calculate

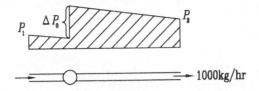

**Fig. 14.10**

the power required by the pump, ignoring any change of height and assuming the skin friction coefficient in the pipe to be 0.01. Take the density of water as $1000\,kg/m^3$.

*Solution*

The power required by the pump can be found using equation (14.12):

$$W = \frac{\dot{m}}{\rho}\Delta P_0$$

The change of height is taken as zero and the change of total pressure is

$$\Delta P_0 = P_1 - P_2 - \Delta P$$

where

$$P_1 = 100 \times 10^3\,\text{Pa}$$
$$P_2 = 500 \times 10^3\,\text{Pa}$$

The pressure drop in the pipe can be found using equation (14.7):

$$\Delta P = 2f\frac{l}{d}\rho\bar{v}^2$$

To find the mean velocity it is necessary to use the continuity equation (13.1):

$$\dot{m} = \rho A\bar{v}$$

$$\frac{1000}{3600} = 1000 \times \frac{\pi}{4}(0.025)^2\bar{v}$$

so that $\bar{v} = 0.566\,\text{m/s}$.

Substituting in equation (14.7)

$$\Delta P = 2 \times 0.01 \times \frac{10}{0.025} \times 1000 \times (0.566)^2$$

$$= 2562.8\,\text{Pa}$$

Therefore

$$\Delta P_0 = 100 \times 10^3 - 500 \times 10^3 - 2562.8$$
$$= -402562.8\,\text{Pa}$$

The pump power is

$$W = \frac{\dot{m}}{\rho}\Delta P_0$$

$$= \frac{1000 \times -402562.8}{3600 \times 1000} = -111.8\,\text{W}$$

*Note:* the negative sign indicates a power input to the pump.

**Example 14.7**

In a hydroelectric system water flows from a reservoir, through a 500 m length of 0.5 m diameter pipe at a mean velocity of 2 m/s to a turbine (Figure 14.11). If the outlet of the turbine is situated 250 m below the free surface of the water in the reservoir, calculate the power output of the system. Assume the only losses are due to friction in the pipe, with $f = 0.008$. Take the density of water as 1000 kg/m³ and $g = 9.81$ m/s².

*Solution*

The pressure drop in the pipe can be found using equation (14.7):

$$\Delta P = 2f\frac{l}{d}\rho\bar{v}^2$$

$$= 2 \times 0.008 \times \frac{500}{0.5} \times 1000 \times (2)^2$$

$$= 64\,000\,\text{Pa} = 64\,\text{kPa}$$

The mass flow rate is given by equation (13.1):

$$\dot{m} = \rho A \bar{v}$$

$$= 1000 \times \frac{\pi}{4}(0.5)^2 \times 2 = 392.7\,\text{kg/s}$$

Assuming the static pressure at the inlet and oulet of the system is equal to the atmospheric pressure:

$$P_1 = P_2 = P_\text{atm}$$

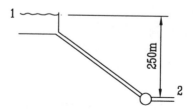

**Fig. 14.11**

the change of total pressure is equal to the difference in hydrostatic pressure less the pressure drop in the pipe.

$$\Delta P_0 = -\Delta P + \rho g(z_1 - z_2)$$
$$= -64\,000 + 1000 \times 9.81 \times 250$$
$$= 2\,388\,500\,\text{Pa} = 2.39\,\text{MPa}$$

Substituting in equation (14.12)

$$W = \frac{\dot{m}}{\rho}\Delta P_0$$

$$= \frac{392.7}{1000} \times 2\,388\,500 = 937\,964\,\text{W} = 938\,\text{kW}$$

*Note:* the power is positive as it is a power output from the system. Compare this answer with that for Example 13.8.

## 14.9 SUMMARY

Key equations that have been introduced in this chapter are as follows.
   Relationship between viscosity and velocity gradient:

$$\tau = \mu\left(\frac{v}{y}\right) \tag{14.2}$$

Torque to overcome friction in a journal bearing:

$$T = \frac{\mu\pi^2 d^3 lN}{2y} \tag{14.3}$$

Reynolds number for flow in a pipe:

$$\text{Re} = \frac{\rho\bar{v}d}{\mu} \tag{14.5}$$

Pressure drop in a pipe:

$$\Delta P = 2f\frac{l}{d}\rho\bar{v}^2 \tag{14.7}$$

Skin friction coefficient for laminar flow in a pipe:

$$f = \frac{16}{\text{Re}} \tag{14.8}$$

Skin friction coefficient for turbulent flow in a smooth pipe:

$$f = 0.08\,\text{Re}^{-1/4} \tag{14.9}$$

Power for flow through a pipeline system:

$$W = \frac{\dot{m}}{\rho}\Delta P_0 \tag{14.12}$$

### 14.10 PROBLEMS

For problems 7 to 10 assume the properties of water to be: $\rho = 1000\,kg/m^3$ and $\mu = 10^{-3}\,kg/m\,s$.

1. A flat slide valve with a face 50 mm wide and 200 mm long moves along a parallel surface with a velocity of 20 m/s. If the two surfaces are separated by an oil film 1 mm thick, find the force on the slide valve. Take the oil to have a viscosity of 0.06 kg/m s.

2. A shaft of 30 mm diameter rotates at 1200 rev/min in a journal bearing. If the bearing is 15 mm wide and the annular clearance between the shaft and bearing is 0.05 mm, calculate the power dissipated at the bearing. Take the viscosity of the lubricating oil to be 0.05 kg/m s.

3. Find the Reynolds number for water flowing through a pipe of 75 mm diameter at 1 m/s. Take the properties of water as: density = $1000\,kg/m^3$; viscosity = $10^{-3}\,kg/m\,s$.

4. A fluid flows through a 25 mm diameter pipe at 4 m/s. Will the flow be laminar, transitional or turbulent if the fluid is:

   (a) oil with $\rho = 900\,kg/m^3$, $\mu = 0.1\,kg/m\,s$;
   (b) steam with $\rho = 2.5\,kg/m^3$, $\mu = 1.2 \times 10^{-5}\,kg/m\,s$;
   (c) sulphuric acid with $\rho = 1800\,kg/m^3$, $\mu = 0.05\,kg/m\,s$.

5. Water is supplied through a pipe of 150 mm diameter and 3 km in length. If the water velocity is 0.5 m/s, calculate the pressure drop in the pipe due to friction. Take the skin friction coefficients as 0.01 and the density of water as $1000\,kg/m^3$.

6. Crude oil flows through a pipeline of 400 mm diameter and 10 km in length. If the oil velocity is 0.4 m/s, find the pressure drop in the pipe. Take the viscosity of the oil as 0.08 kg/m s. Assume the flow to be laminar.

7. In a factory water flows through a smooth-bore pipe of 15 mm diameter at a velocity of 1 m/s. If the pipe is horizontal and is 100 m long, calculate the head required to overcome friction.

8. Water flows through a smooth-bore pipe with a total length of 1200 m. The first 500 m has a diameter of 0.5 m, the remaining pipe has a diameter of 0.8 m. If the velocity in the smaller bore is 1.2 m/s, calculate the total pressure drop for the system.

9. Water is fed to a header tank situated 10 m above a pump, through a smooth pipe of 12 mm diameter and 35 m length. What power is required to pump the water at a velocity of 1.5 m/s through the pipe?

10. In a hydroelectric system water flows from a reservoir through a 0.8 m diameter pipe at a velocity of 1.5 m/s. If the outlet of the turbine is situated 300 m below the free surface and the pipe is 800 m long, calculate the power output of the system. Assume the only loss to be due to friction and the pipe to have a smooth bore.

# Basic heat transfer 15

## 15.1 AIMS

- To introduce the modes of heat transfer: convection, conduction and radiation.
- To discuss radiation between two parallel surfaces.
- To define the overall heat transfer coefficient for combined modes of heat transfer.
- To evaluate the overall heat transfer coefficient using a thermal resistance analogy.

## 15.2 MODES OF HEAT TRANSFER

Heat is a form of energy that is transferred across the boundary of a thermofluid system as a result of a temperature difference. The greater the temperature difference the more rapidly will the heat be transferred. Conversely, the lower the temperature difference, the slower will be the rate at which heat is transferred. When discussing the modes of heat transfer it is the rate of heat transfer, $Q$, that defines the situation rather than the quantity of heat.

There are three distinct modes of heat transfer – convection, conduction and radiation. Although two, or even all three, modes of heat transfer may be combined in any particular thermofluid situation, the three can be analysed in a different way and will be introduced separately.

### 15.2.1 Convection

Convection is a mode of heat transfer that takes place due to motion within a fluid. In Chapter 14 it was shown that motion of a fluid is either laminar or turbulent. In the case of turbulent flow, particles can move at random within the flow. Consider a particle moving normal to a surface from position 1 to position 2, as shown in Figure 15.1.

If the fluid, shown in Figure 15.1, starts at a constant temperature and the surface is suddenly increased in temperature to above that of the fluid, there will be convective heat transfer from the surface to the fluid as a result of the temperature difference. The particle at 1 starts in contact with the surface and its temperature will rise as a result of that contact. If the particle then moves to position 2, it will have a higher temperature

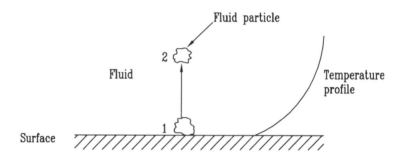

**Fig. 15.1** Mechanism of convection.

than the fluid around it. As a result, the temperature of the particle will fall as heat is transferred to the fluid around it, until the particle and the surrounding fluid reach thermal equilibrium.

Within this model, heat is transferred from the surface to the fluid in the region of position 2 by means of movement of one fluid particle. In real situations, all the particles within the fluid will be in motion and there will be a gradual drop in temperature from the surface through the subsequent layers of the fluid, as defined by the temperature profile shown in Figure 15.1. Under these conditions the temperature difference causing the heat transfer can be defined as

$\Delta T$ = surface temperature − mean fluid temperature

Using this definition of the temperature difference, the rate of heat transfer due to convection can be evaluated using Newton's law of cooling, as described in section 1.5.3:

$$Q = h_c A \Delta T \qquad (15.1)$$

where $A$ is the heat transfer surface area and $h_c$ is the coefficient of heat transfer from the surface to the fluid, referred to as the 'convective heat transfer coefficient'.

The units of the convective heat transfer coefficient can be determined from the units of the other variables:

$$Q = h_c A \Delta T$$
$$W = (h_c) \, m^2 \, K$$

so the units of $h_c$ are $W/m^2 \, K$.

The relationship given in equation (15.1) is true irrespective of whether the flow is laminar or turbulent. It is also true for the situation where a surface is being heated due to the fluid having a higher temperature than the surface. However, in this case the direction of heat transfer is from the fluid to the surface and the temperature difference will be

$\Delta T$ = mean fluid temperature − surface temperature

The relative temperature of the surface and fluid determine the direction of heat transfer and the rate at which it takes place. Figure 15.2 illustrates the two alternative convective situations of heating and cooling. Figure

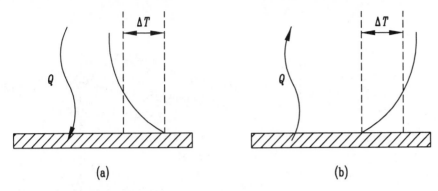

**Fig. 15.2** Convective heat transfer situations.

15.2(a) defines the situation in which a surface is being 'heated' by a fluid at a higher temperature, so that there is a temperature drop from the fluid to the surface. Figure 15.2(b) defines the situation in which the surface is being cooled by a fluid at a lower temperature, so that there is a temperature drop from the surface to the fluid.

As given in equation (15.1), the rate of heat transfer is determined not only by the temperature difference but also by the convective heat transfer coefficient $h_c$. This is not a constant but varies quite widely, depending on the properties of the fluid and the behaviour of the flow. It is possible to get an overall view of the convective properties of fluids by considering the simple model defined in Figure 15.1. This indicates that the value of $h_c$ depends on the 'thermal capacity' of the fluid particle considered.

The thermal capacity of a particle of given volume is proportional to $\rho C_p$. In other words, the higher the density and specific heat of the fluid, the higher the convective heat transfer coefficient. Comparing two common heat transfer fluids, namely air and water, water is approximately 800 times more dense than air and also has a higher value of specific heat. If the argument given above is valid, then water has a higher thermal capacity than air and should have a better convective heat transfer performance. This is borne out in practice because typical values of convective heat transfer coefficients are as shown in Table 15.1. The variation in the values reflects the variation in the behaviour of the flow, particularly the flow velocity, with the higher values of $h_c$ resulting from higher flow velocities over the surface.

**Table 15.1**

| Fluid | $h_c \, (\mathrm{W/m^2\,K})$ |
|-------|------------------------------|
| Water | 500–10 000 |
| Air | 5–1000 |

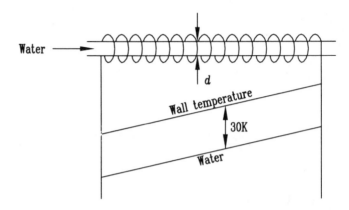

**Fig. 15.3**

### Example 15.1

In an electric water heater (Figure 15.3), 1.0 litre/min of water is raised from 20 to 60 °C. The water flows through a single pipe of 8 mm diameter. If there is no heat loss from the outside of the pipe and the convective heat transfer coefficient on the water side is 2000 W/m² K, find the required length of pipe.

Assume that the temperature difference between the pipe and water is 30 K throughout the whole length. Take $C_P$ for water as 4.2 kJ/kg K.

*Solution*

The rate of heat transfer, $Q$, can be evaluated from the steady flow rate equation defined in equation (13.9):

$$Q = \dot{m}(h_2 - h_1)$$
$$= \dot{m}C(T_2 - T_1)$$
$$= \frac{1}{60} \times 4200\,(60 - 20)$$
$$= 2800\,\text{W} = 2.8\,\text{kW}$$

From equation (15.1):

$$Q = h_c A \Delta T$$

and

$$A = Q/h_c \Delta T = 2800/(2000 \times 30) = 0.0467\,\text{m}^2$$

Now

$$A = \pi d l$$

so that

$$l = \frac{A}{\pi d} = \frac{0.0467}{\pi \times 0.008} = 1.86\,\text{m}$$

### 15.2.2 Conduction

If a fluid could be kept stationary, no convection would take place. However, it would still be possible to transfer heat within the fluid by means of conduction. Conduction depends on the transfer of energy from one molecule to another within the heat transfer medium and, in this sense, thermal conduction is analogous to electrical conduction.

Conduction can occur within both solids and fluids. The rate of heat transfer depends on a physical property of the particular solid or fluid, termed its thermal conductivity $k$, and the temperature gradient across the medium. The thermal conductivity is defined as the measure of the rate of heat transfer across a unit width of material, for a unit cross-sectional area and for a unit difference in temperature.

Figure 15.4 defines a conduction situation in which heat transfer is taking place across a solid wall. In more complex situations it is possible to have conduction taking place simultaneously in three directions, but for the present discussion it is sufficient to consider conduction in just one direction.

From the definition of thermal conductivity $k$ it can be shown that the rate of heat transfer is given by the relationship

$$Q = \frac{kA\Delta T}{x} \tag{15.2}$$

where $\Delta T$ is the temperature difference $T_1 - T_2$ defined by the temperatures on either side of the wall.

The units of thermal conductivity can be determined from the units of the other variables:

$$Q = kA\Delta T/x$$

$$W = (k)\,m^2\,K/m$$

so the units of $k$ are $W/m^2\,K/m$, expressed as $W/m\,K$.

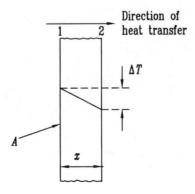

**Fig. 15.4** Conduction in one direction.

**Table 15.2** Thermal conductivities for some common materials

| Material | Thermal conductivity (W/m K) |
|---|---|
| Metals | |
|   aluminium | 210 |
|   copper | 360 |
|   steel | 44 |
| Building materials | |
|   brick | 0.7 |
|   concrete | 0.8 |
|   wood | 0.2 |
| Thermal insulation | |
|   asbestos | 0.1 |
|   foam plastic | 0.04 |
|   glass fibre | 0.05 |

Since thermal conductivity is analogous to electrical conductivity, it would follow that metals, having good electrical conductivity, should have good thermal conductivity. Similarly, electrical insulators should have low thermal conductivity. This is borne out in Table 15.2.

**Example 15.2**

An oven has a brick wall of 80 mm thickness and thermal conductivity of 1 W/m K (Figure 15.5). The convective heat transfer coefficient between the outer surface of the wall and the surrounding air is 10 W/m² K. If the inner surface of the oven wall is at 400 °C and the outside air is at 20 °C, calculate:

(a) the outside surface temperature of the wall,
(b) the rate of heat transfer, per unit area, through the wall.

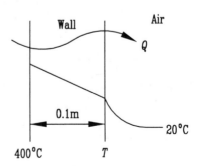

**Fig. 15.5**

*Solution*

Heat transfer across the wall is by conduction. From equation (15.2)

$$Q = \frac{kA\Delta T}{x} = \frac{1(400 - T)A}{0.08}$$

Therefore

$$\frac{Q}{A} = 5000 - 12.5\,T \qquad\qquad (15.2a)$$

Heat transfer from the wall to the air is by convection. From equation (15.1)

$$Q = h_c A\Delta T = 10(T - 20)A$$

Therefore

$$\frac{Q}{A} = 10T - 200 \qquad\qquad (15.2b)$$

Assuming $Q/A$ to be the same for both modes of heat transfer, combining equations (15.2a) and (15.2b) gives

$$5000 - 12.5T = 10T - 200$$

Solving this equation gives the solution to part (a):

$$T = \frac{5200}{22.5} = 231.1\,°C$$

Substituting in equation (15.2a) gives the solution to part (b):

$$Q/A = 5000 - 12.5\,(231.1) = 2111.3\,W/m^2$$

### 15.2.3  Radiation

The third mode of heat transfer, radiation, does not depend on any medium for its transmission. In fact, it takes place most readily when there is a perfect vacuum between the emitter and receiver of such energy. This is proved on a daily basis by the transfer of energy from the sun to the earth through space.

Radiation is a form of electromagnetic energy transmission and takes place between all matter providing that it is at a temperature above absolute zero. Infra-red, visible light and ultraviolet radiation form just part of the overall electromagnetic spectrum. Radiation is energy emitted by the electrons vibrating in the molecules at the surface of a body. The amount of energy that can be transferred depends on the absolute temperature of the body and the radiant properties of the surface.

A body that has a surface that will absorb all the radiant energy it receives is an ideal radiator, termed a **black body**. Such a body will not only absorb radiation at a maximum level but will also emit radiation at a maximum level.

The rate of heat transfer due to radiation from the surface of a black body can be evaluated using an equation derived by Stefan and Boltzmann in the latter part of the nineteenth century:

$$Q = \sigma A T^4 \qquad (15.3)$$

where $\sigma$ is the Stefan–Boltzmann constant, $5.67 \times 10^{-8} \, \text{W/m}^2 \, \text{K}^4$.

Equation (15.3) shows that any black body surface above a temperature of absolute zero will radiate heat at a rate that is proportional to the fourth power of the absolute temperature. While the rate of heat transfer from a black body is independent of the condition of the surroundings, most practical situations involve radiation between two surfaces in which it is the net exchange of heat transfer that is important.

Such a situation is shown in Figure 15.6. If the two parallel surfaces are considered to be large compared with the distance between them, the losses around the edges can be considered to be negligible. The rate of heat transfer from surface 1 will be

$$Q_1 = \sigma A T_1^4$$

Similarly, there will be a rate of heat transfer from surface 2:

$$Q_2 = \sigma A T_2^4$$

It follows that the net rate of heat transfer between the surfaces will be

$$Q = Q_1 - Q_2 = \sigma A (T_1^4 - T_2^4) \qquad (15.4)$$

The situation shown in Figure 15.6 might be considered to be fairly academic. However, in practice, it is typical of the situation found in a domestic electric storage radiator. This consists of a hot core with an outer sheet metal cover. The surface of the hot core and the outer cover act as two parallel plates forming an open channel for air to flow through. Heat transfer from the core takes place by radiation, $Q_R$, between the core and the outer cover together with convection, $Q_C$, between the surface of the core and the air.

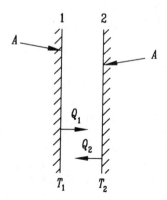

**Fig. 15.6** Radiation between two black body surfaces.

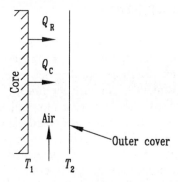

**Fig. 15.7** Heat transfer in a storage radiator.

**Example 15.3**

In a storage radiator the surface of the core has an area of $0.6\,m^2$ and a temperature of 220 °C (Figure 15.7). If the outer cover has the same surface area and a temperature of 65 °C, calculate the rate of heat transfer taking place. Assume the surfaces to act as black bodies and the convective heat transfer coefficient between the surface of the core and the air to be $10\,W/m^2\,K$. Take the ambient temperature to be 20 °C.

*Solution*

The rate of heat transfer is the sum of the individual heat transfers by radiation and convection from the core:

$$Q = Q_R + Q_C$$

From equation (15.4)

$$Q_R = \sigma A(T_1^4 - T_2^4)$$

Now

$$T_1 = 220 + 273 = 493\,K$$
$$T_2 = 65 + 273 = 338\,K$$

and

$$Q_R = 5.67 \times 10^{-8} \times 0.6\,(493^4 - 338^4)$$
$$= 1565.6\,W$$

From equation (15.1)

$$Q_C = h_c A(T_1 - T_a)$$
$$= 10 \times 0.6\,(220 - 20)$$
$$= 1200\,W$$

Therefore

$$Q = 1565.6 + 1200 = 2765.6\,W$$

## 15.3 OVERALL HEAT TRANSFER COEFFICIENT

For steady state heat transfer, the three modes of heat transfer do not occur as isolated events. In fact, for most practical situations, heat transfer relies on two, or even all three, modes occurring together. For such situations, it is inconvenient to analyse each mode separately. Therefore, it is useful to derive an overall heat transfer coefficient that will combine the effect of each mode within a general steady state situation.

A central heating unit in which hot water flows through a heat exchanger and so heats the air in a room, is termed a 'radiator'. Similarly, an air-cooled heat exchanger, necessary to cool the hot water from a car engine is also termed a 'radiator'. With the widespread use of such a term it would be imagined that radiation is the most important mode of heat transfer. However, this is not the case. Both types of radiator described above rely on a combination of convection and conduction, with radiation playing only a small part in the operation.

In the case of a central-heating radiator, the heat transfer processes involve convection from the hot water to the inner surface of the heat exchanger, conduction across the metal wall and convection from the outer surface to the surrounding air. These combined modes of convection and conduction are shown in Figure 15.8.

Assuming the convective heat transfer coefficients for water and air are $h_{c_w}$ and $h_{c_a}$ respectively, and the thermal conductivity of the wall is $k$, the situation defined in Figure 15.8 can be analysed as follows.

Water-side – from equation (15.1)

$$Q = h_{c_w} A(T_1 - T_2)$$

Therefore

$$T_1 - T_2 = \frac{Q}{A h_{c_w}}$$

Across the wall – from equation (15.2)

$$Q = \frac{k}{x} A(T_2 - T_3)$$

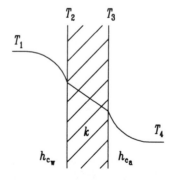

**Fig. 15.8** Combined convection and conduction.

Therefore

$$T_2 - T_3 = \frac{Qx}{Ak}$$

Air-side – from equation (15.1)

$$Q = h_{c_a} A(T_3 - T_4)$$

Therefore

$$T_3 - T_4 = \frac{Q}{Ah_{c_a}}$$

Assuming that the rate of heat transfer $Q$ is the same across the combined situation, the temperature differences can be summed to give

$$(T_1 - T_2) + (T_2 - T_3) + (T_3 - T_4) = \frac{Q}{Ah_{c_w}} + \frac{Qx}{Ak} + \frac{Q}{Ah_{c_a}}$$

resulting in

$$T_1 - T_4 = \frac{Q}{A}\left(\frac{1}{h_{c_w}} + \frac{x}{k} + \frac{1}{h_{c_a}}\right) \tag{15.5}$$

The temperature difference $(T_1 - T_4)$ represents the difference for the overall situation and can be expressed as $\Delta T$. Similarly, the surface heat transfer coefficients and the thermal conduction across the wall can be combined into an overall heat transfer coefficient $U$, defined by

$$\frac{1}{U} = \frac{1}{h_{c_w}} + \frac{x}{k} + \frac{1}{h_{c_a}} \tag{15.6}$$

Combining equations (15.5) and (15.6) gives a general relationship

$$Q = UA\Delta T \tag{15.7}$$

This relationship is based upon an **overall heat transfer coefficient** which includes convection on both sides of a plane wall and conduction across the wall. By using similar models to that given in Figure 15.8, the overall heat transfer coefficient can be derived for any combination of convection and conduction.

## Example 15.4

Calculate the overall heat transfer coefficient for a heat exchanger with water and air operating either side of a brass pipe 2 mm thick. Assume convective heat transfer coefficients of 2000 and 50 W/m² K for the water- and air-sides respectively. The thermal conductivity for brass can be taken as 100 W/m K.

Assuming the thickness of the pipe to be small compared with the diameter, Figure 15.8 can be used.

*Solution*

From equation (15.6)

$$\frac{1}{U} = \frac{1}{h_{c_w}} + \frac{x}{k} + \frac{1}{h_{c_a}}$$

$$= \frac{1}{2000} + \frac{0.002}{100} + \frac{1}{50}$$

$$= 0.0005 + 0.00002 + 0.02$$

$$= 0.02052 \, \text{m}^2 \, \text{K/W}$$

Therefore

$$U = 1/0.02052 = 49 \, \text{W/m}^2 \, \text{K}$$

*Note:* the overall heat transfer coefficient in this case is nearly equal to the air-side heat transfer coefficient. In other words, the poor performance on the air-side dominates the overall situation and the good performance on the water-side is not utilized.

In practice, this can be compensated for by increasing the area on the air-side, which is why car radiators have fins on the air-side.

## 15.4   THERMAL RESISTANCE

The geometry described above represents quite a simple heat transfer situation. Not all heat exchangers employ water or air as the working fluids; there are a range of other fluids that can be used, although water and air tend to be used most widely. Similarly, not all heat transfer situations involve a plane wall made of a single material. Many involve composite walls consisting of several materials. For example, a domestic refrigerator has a sheet metal outside casing, a moulded plastic inner shell, with the space between filled with insulation. Taking a cross-section through any of the walls would reveal a composite structure of sheet metal, insulation and sheet plastic bonded together.

To analyse heat transfer situations involving fluids and composite walls requires the evaluation of an overall heat transfer coefficient, but one requiring a rather more complex relationship than that given in equation (15.6). The most straightforward way of evaluating such an overall heat transfer coefficient is by using an electrical resistance analogy for the thermal resistance of each separate mode of heat transfer.

With convection, the rate of heat transfer is a function of $h_c$, the convective heat transfer coefficient. For a given temperature difference between the surface and the fluid, the higher the value of $h_c$ the higher the rate of heat transfer. Alternatively, it could be reasoned the higher the value of $h_c$ the lower the 'resistance' to heat transfer. It is, therefore, possible to define the thermal resistance of a convective process as

$$R \, (\text{convection}) = \frac{1}{h_c}$$

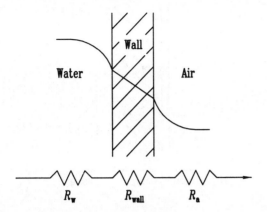

**Fig. 15.9** Resistance analogy for heat transfer.

The thermal resistance for conduction can be thought of as reducing the higher the thermal conductivity becomes, but increasing directly with the thickness of the material:

$$R \text{ (conduction)} = \frac{x}{k}$$

Using these definitions of thermal resistance, equation (15.6) for the overall heat transfer coefficient can be expressed in the form

$$\frac{1}{U} = \frac{1}{h_{c_w}} + \frac{x}{k} + \frac{1}{h_{c_a}}$$

$$R_0 = R_w + R_{\text{wall}} + R_a$$

where $R_0$ is the overall thermal resistance equivalent to $1/U$. The overall thermal resistance can be found by summing the individual thermal resistances in the same way as electrical resistances operating in series. The combined convection and conduction situation, shown in Figure 15.8, can now be defined in terms of the individual thermal resistance, as given in Figure 15.9.

Clearly, the resistance analogy can be applied to any number of thermal resistances in series, such that

$$R_0 = R_1 + R_2 + R_3 \dots \tag{15.8}$$

**Example 15.5**

A domestic refrigerator has a heat transfer surface area of $2\,\mathrm{m}^2$ (Figure 15.10). If the temperature inside the refrigerator is maintained at $3\,^{\circ}\mathrm{C}$ when the surrounding air is at $20\,^{\circ}\mathrm{C}$, calculate the rate of heat gain by the refrigerator.

Assume the walls of the refrigerator to consist of an outside steel skin of 2 mm thickness, insulation of 25 mm thickness and an inner plastic casing

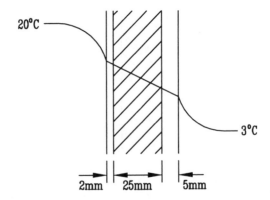

**Fig. 15.10**

of 5 mm thickness. Take the values of thermal conductivity to be (W/m K): $k$ (steel) 40; $k$ (insulation) 0.05; $k$ (plastic) 1. Take the convective heat transfer coefficients on the inner and outer surfaces to be 10 W/m² K.

*Solution*

The overall heat transfer coefficient can be evaluated using the resistance analogy defined by equation (15.8):

$$R_0 = R_a + R_s + R_i + R_p + R_a$$

where

$$R_a = 1/h_{c_a} = 1/10 = 0.1$$
$$R_s = (x/k)_s = 0.002/40 = 0.000\,05$$
$$R_i = (x/k)_i = 0.025/0.05 = 0.5$$
$$R_p = (x/k)_p = 0.005/1 = 0.005$$
$$R_a = 1/h_{c_a} = 1/10 = 0.1$$

Therefore

$$R_0 = 0.705\,05 \text{ m}^2 \text{ K/W}$$

and

$$U = \frac{1}{R_0} = \frac{1}{0.705} = 1.42 \text{ W/m}^2 \text{ K}$$

Substituting in equation (15.7)

$$Q = UA\Delta T = 1.42 \times 2 \times (20 - 3)$$
$$= 48.2 \text{ W.}$$

## 15.5 SUMMARY

Key equations that have been introduced in this chapter are as follows.
For convection between a surface and an adjacent fluid:

$$Q = h_c A \Delta T \qquad (15.1)$$

For conduction across a wall:

$$Q = \frac{k}{x} A \Delta T \qquad (15.2)$$

For radiation between two black body surfaces:

$$Q = \sigma A (T_1^4 - T_2^4) \qquad (15.4)$$

For combined convection and conduction for water and air across a plane wall:

$$\frac{1}{U} = \frac{1}{h_{c_w}} + \frac{x}{k} + \frac{1}{h_{c_a}} \qquad (15.6)$$

General relationship for combined convection and conduction:

$$Q = U A \Delta T \qquad (15.7)$$

Thermal resistances in series:

$$R_0 = R_1 + R_2 + R_3 \ldots \qquad (15.8)$$

## 15.6 PROBLEMS

1. An electronic device dissipates 200 W with a maximum surface temperature of 100 °C. Find the surface area of heat sink required to cool the device if the heat transfer coefficient on the air-side is 15 W/m² K and the maximum ambient temperature is 30 °C.
2. A hair dryer incorporates a 750 W electrical resistance heater. If the heater consists of 0.8 mm diameter wire and has a maximum temperature of 400 °C, find the length of wire required if the air is heated from 20 to 60 °C and the heat transfer coefficient for air flowing over the wire is 30 W/m² K.
3. An electric storage heater dissipates 1.5 kW with a core surface temperature of 150 °C and a temperature of the outer cover of 70 °C. Estimate the required surface area of the core if the air-side heat transfer coefficient is 12 W/m² K. Take the temperature of the air in the room to be 25 °C.
4. A furnace has a refractory wall 0.128 m thick with a thermal conductivity of 0.8 W/m K. If the inside surface temperature is 700 °C and the outside surface temperature is at 300 °C, calculate:

(a) the rate of heat transfer per unit of wall;
(b) the outside heat transfer coefficient if the air outside is at 25 °C.

5. The furnace wall defined in problem 4 is insulated on the outside by a 25 mm layer of fibreglass, protected by 12 mm sheet steel. If the thermal conductivities are 0.05 W/m K for fibreglass and 40 W/m K for steel find:

   (a) the new rate of heat transfer;
   (b) the outside temperature of the steel.

6. A window $0.25 \times 0.2$ m is situated in the door of an electric oven. If the window consists of two layers of glass, 7 mm thick, with a 5 mm air gap between, calculate the heat loss through the window when the oven is operating at 200 °C. Assume an outside temperature of 25 °C. Take thermal conductivities of 1 W/m K for glass and 0.026 W/m K for the air gap, with heat transfer coefficients of 10 W/m$^2$ K both inside and outside.

7. An electric kettle can be modelled as a plastic cylinder 0.15 m in diameter and 0.2 m high. The plastic can be taken as being 2 mm thick with a thermal conductivity of 1 W/m K. Calculate the heat loss from the kettle when the water is at 100 °C and the surrounding air is at 20 °C. The heat transfer coefficient on the water-side can be taken as 3000 W/m$^2$ K and on the air-side as 10 W/m$^2$ K.

8. A domestic freezer has an internal capacity of $1 \times 0.5 \times 0.5$ m. The temperature inside is maintained at $-5$ °C in an ambient temperature of 35 °C. Calculate the rate of heat transfer to the freezer if the walls, and door, consist of: 5 mm plastic, $k = 1$ W/m K; 15 mm insulation, $k = 0.05$ W/m K; 3 mm steel, $k = 40$ W/m K; and the heat transfer coefficient is 8 W/m$^2$ K on both sides of the freezer.

# Appendix A　Mechanical properties of metals

$E$ = modulus of elasticity (GPa)

$G$ = shear modulus (GPa)

$v$ = Poisson's ratio

$\sigma_y$ = yield stress (MPa)

$\sigma_u$ = ultimate tensile stress (MPa)

$\rho$ = density (kg/m³)

| | $E$ | $G$ | $v$ | $\sigma_y$ | $\sigma_u$ | $\rho$ |
|---|---|---|---|---|---|---|
| *Pure metals* | | | | | | |
| Aluminium | 70 | 25 | 0.34 | 20 | 50 | 2 700 |
| Copper | 130 | 48 | 0.34 | 33 | 210 | 8 960 |
| Gold | 79 | 26 | 0.42 | | 100 | 19 300 |
| Iron | 210 | 81 | 0.29 | 100 | 350 | 7 900 |
| Lead | 16 | 6 | 0.44 | | 12 | 11 300 |
| Nickel | 200 | 76 | 0.31 | 60 | 310 | 8 900 |
| Tin | 47 | 17 | 0.36 | 14 | 200 | 7 300 |
| Titanium | 120 | 46 | 0.36 | 200 | 350 | 4 500 |
| | | | | | | |
| *Alloys* | | | | | | |
| Aluminium 2024 (4.5% Cu) | 72 | 28 | 0.33 | 395 | 475 | 2 800 |
| Brass (70/30) | 101 | 37 | 0.35 | 115 | 320 | 8 550 |
| Cast iron | 100 | 40 | 0.26 | 100 | 150 | 7 000 |
| Magnesium alloy, AZM | 45 | 17 | 0.35 | 160 | 230 | 1 800 |
| Nimonic 80A (super alloy) | 214 | 80 | 0.35 | 800 | 1300 | 8 190 |
| Phosphor-bronze (5% Sn) | 100 | 36 | 0.38 | 390 | 540 | 8 850 |
| Soft solder (50% Sn) | 40 | | | 33 | 42 | 8 900 |
| Steel: mild | 210 | 81 | 0.29 | 240 | 320 | 7 850 |
| Steel: medium carbon | 210 | 81 | 0.30 | 320 | 400 | 7 850 |
| Steel: stainless | 200 | 78 | 0.29 | 255 | 660 | 7 900 |
| Titanium 6Al-4V | 115 | 43 | 0.33 | 850 | 950 | 4 420 |

# Appendix B Second moment of area of a rectangle

Consider the rectangle with breadth, $b$, and depth, $d$, as shown in Figure B.1. In order to find the second moment of area for the rectangle about axis X–X, it is necessary first to consider the elemental strip shown.

$$\text{second moment of area for the elemental strip} = b\,dy(y+h)^2$$
$$= b(y^2 + 2yh + h^2)\,dy$$

Integrating for the whole rectangle between the limits of $d/2$ and $-d/2$:

$$I_{xx} = \int b(y^2 + 2yh + h^2)\,dy$$
$$= \int by^2\,dy + h\int 2y\,dy + h^2\int b\,dy$$
$$= \frac{bd^3}{12} + h(0) + h^2(bd)$$

where $bd^3/12$ is the second moment of area about the centroid of the rectangle and $bd$ is the area of the rectangle. Therefore

$$I_{xx} = \frac{bd^3}{12} + Ah^2$$

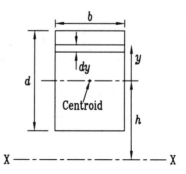

Fig. B.1

# Appendix C  Saturated water–steam properties

| Pressure, $P$ (kPa) | Temperature, $T_s$ (°C) | Density, $\rho_g$ (kg/m³) | Internal energy (kJ/kg) | | Enthalpy (kJ/kg) | |
|---|---|---|---|---|---|---|
| | | | $u_f$ | $u_g$ | $h_f$ | $h_g$ |
| 10 | 45.8 | 0.068 | 192 | 2438 | 192 | 2585, |
| 20 | 60.1 | 0.131 | 252 | 2457 | 252 | 2610 |
| 30 | 69.1 | 0.191 | 289 | 2469 | 289 | 2625 |
| 40 | 75.9 | 0.250 | 318 | 2477 | 318 | 2637 |
| 50 | 81.4 | 0.309 | 341 | 2484 | 341 | 2646 |
| 60 | 86.0 | 0.366 | 360 | 2490 | 360 | 2654 |
| 80 | 93.5 | 0.479 | 392 | 2499 | 392 | 2666 |
| 100 | 99.6 | 0.590 | 417 | 2506 | 418 | 2675 |
| 200 | 120.2 | 1.129 | 505 | 2529 | 505 | 2706 |
| 300 | 133.5 | 1.651 | 561 | 2543 | 561 | 2725 |
| 400 | 143.6 | 2.164 | 604 | 2553 | 605 | 2738 |
| 500 | 151.9 | 2.669 | 640 | 2560 | 640 | 2748 |
| 600 | 158.8 | 3.170 | 670 | 2566 | 670 | 2756 |
| 800 | 170.4 | 4.162 | 720 | 2575 | 721 | 2768 |
| 1000 | 179.9 | 5.147 | 762 | 2581 | 763 | 2776 |
| 1500 | 198.3 | 7.595 | 843 | 2592 | 845 | 2790 |
| 2000 | 212.4 | 10.045 | 906 | 2598 | 909 | 2797 |
| 3000 | 233.8 | 15.008 | 1005 | 2602 | 1008 | 2802 |
| 4000 | 250.3 | 20.100 | 1082 | 2601 | 1087 | 2800 |

# Appendix D Superheated steam properties

| Pressure, $P$ (kPa) | Temperature $T_s$ (°C) | | Steam temperature (°C) | | | | | |
|---|---|---|---|---|---|---|---|---|
| | | | 150 | 200 | 250 | 300 | 400 | 500 |
| 100 | 99.6 | $\rho$ | 0.516 | 0.460 | 0.416 | 0.379 | 0.322 | 0.280 |
| | | $u$ | 2583 | 2658 | 2734 | 2811 | 2968 | 3132 |
| | | $h$ | 2776 | 2875 | 2975 | 3075 | 3278 | 3488 |
| 200 | 120.2 | $\rho$ | 1.042 | 0.926 | 0.834 | 0.760 | 0.645 | 0.561 |
| | | $u$ | 2577 | 2654 | 2731 | 2809 | 2967 | 3131 |
| | | $h$ | 2769 | 2871 | 2971 | 3072 | 3277 | 3487 |
| 300 | 133.5 | $\rho$ | 1.578 | 1.396 | 1.256 | 1.142 | 0.970 | 0.843 |
| | | $u$ | 2570 | 2651 | 2729 | 2807 | 2966 | 3130 |
| | | $h$ | 2760 | 2866 | 2968 | 3070 | 3275 | 3486 |
| 400 | 143.6 | $\rho$ | 2.125 | 1.872 | 1.680 | 1.527 | 1.294 | 1.125 |
| | | $u$ | 2564 | 2647 | 2726 | 2805 | 2965 | 3129 |
| | | $h$ | 2752 | 2860 | 2965 | 3067 | 3274 | 3485 |
| 500 | 151.9 | $\rho$ | | 2.353 | 2.108 | 1.914 | 1.620 | 1.407 |
| | | $u$ | | 2643 | 2724 | 2804 | 2964 | 3128 |
| | | $h$ | | 2855 | 2961 | 3065 | 3272 | 3484 |
| 600 | 158.8 | $\rho$ | | 2.841 | 2.539 | 2.302 | 1.947 | 1.690 |
| | | $u$ | | 2639 | 2721 | 2802 | 2962 | 3128 |
| | | $h$ | | 2850 | 2958 | 3062 | 3271 | 3483 |
| 800 | 170.4 | $\rho$ | | 3.835 | 3.411 | 3.085 | 2.603 | 2.256 |
| | | $u$ | | 2630 | 2716 | 2798 | 2960 | 3126 |
| | | $h$ | | 2839 | 2950 | 3057 | 3268 | 3481 |
| 1000 | 179.9 | $\rho$ | | 4.856 | 4.296 | 3.876 | 3.263 | 2.825 |
| | | $u$ | | 2621 | 2710 | 2794 | 2958 | 3124 |
| | | $h$ | | 2827 | 2943 | 3052 | 3264 | 3478 |
| 2000 | 212.4 | $\rho$ | | | 8.973 | 7.968 | 6.617 | 5.696 |
| | | $u$ | | | 2680 | 2774 | 2946 | 3116 |
| | | $h$ | | | 2902 | 3025 | 3249 | 3467 |
| 3000 | 233.8 | $\rho$ | | | 14.174 | 12.321 | 10.069 | 8.615 |
| | | $u$ | | | 2643 | 2752 | 2935 | 3108 |
| | | $h$ | | | 2855 | 2995 | 3233 | 3456 |
| 4000 | 250.3 | $\rho$ | | | | 16.997 | 13.628 | 11.582 |
| | | $u$ | | | | 2727 | 2922 | 3100 |
| | | $h$ | | | | 2962 | 3216 | 3445 |

Units: $\rho$ in kg/m$^3$, $u$ in kJ/kg, $h$ in kJ/kg.

# Appendix E    Answers

CHAPTER   1

1.

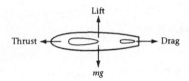

2. Yes, an open system in which both heat and work cross the boundary.

3. $J = N \times m = \dfrac{kg\,m}{s^2}\,m = \dfrac{kg\,m^2}{s^2}$

4. $3\,h\,21\,min = 201\,min = 120\,60\,s = 12.06\,ks$

5. $W = 300 \times 9.81 \times \dfrac{50}{10} = 14\,715\,W = 14.7\,KW$

6. $37\,°C = 37 + 273 = 310\,K$

CHAPTER   2

1.

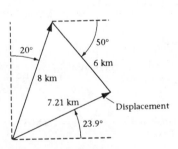

2. $V_1 = \dfrac{54\,000}{3600} = 15\,m/s,\ s_A = \dfrac{(0+15)200}{2} = 1500\,m,\ s_B = \dfrac{(15+15)300}{2} = 4500\,m,$

$s_C = \dfrac{(15+0)150}{2} = 1125\,m$

Total $s = 1500 + 4500 + 1125 = 7125\,m$ or $7.125\,km$

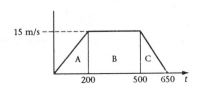

3. $s_A = s_C = \dfrac{v_2 t_A}{2} = 0.15\,\text{m}$ $\therefore t_A = t_C = \dfrac{0.3}{v_2}$

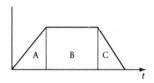

$s_B = v_2 t_B = 0.3\,\text{m}$ $\therefore t_B = \dfrac{0.3}{v_2}$, $t = 1.2\,\text{s} = \dfrac{0.3}{v_2} + \dfrac{0.3}{v_2} + \dfrac{0.3}{v_2} = \dfrac{0.9}{v_2}$

$\therefore v_2 = \dfrac{0.9}{1.2} = 0.75\,\text{m/s}$

4.

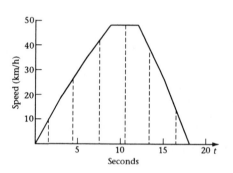

(a) $v_{av} = \dfrac{10 + 26 + 41 + 48 + 37 + 14}{6} = 29.3\,\text{km/h}$ or $\dfrac{29\,300}{3600} = 8.14\,\text{m/s}$

$s = vt = 8.15 \times 18 = 147\,\text{m}$

(b) $v = \dfrac{19\,000}{3600} = 5.28\,\text{m/s}$, $a = \dfrac{v}{t} = \dfrac{5.28}{3} = 1.76\,\text{m/s}^2$

(c) $v_0 = \dfrac{27\,000}{3600} = 7.5\,\text{m/s}$, $a = \dfrac{-v_0}{t} = \dfrac{7.5}{3} = 2.5\,\text{m/s}^2$

(d) $v_{av} = 29.3\,\text{km/h}$.

5. $v_1 = \sqrt{(4^2 + 2 \times 1.64 \times 5)} = 5.7\,\text{m/s}$

6. (a) $t = \sqrt{\dfrac{2 \times 60}{9.81}} = 3.5\,\text{s}$,

(b) $s = \left(\dfrac{2v}{2}\right) t = \dfrac{(2 \times 120)}{2} 3.5 = 420\,\text{m}$

7. (a) $\omega_0 = 2\pi\,\text{rad/s}$, $\omega_1 = 16\pi\,\text{rad/s}$, $\alpha = \dfrac{16\pi - 2\pi}{20} = 2.2\,\text{rad/s}^2$

(b) $a = 2.2 \times 0.35 = 0.77\,\text{m/s}^2$

8. (a) $\omega_0 = \dfrac{300 \times 2\pi}{60} = 31.4 \,\text{rad/s}, \quad \omega_1 = \dfrac{360 \times 2\pi}{60} = 37.7 \,\text{rad/s},$

$\alpha = \dfrac{37.7^2 - 31.4^2}{2(2\pi \times 18)} = 1.9 \,\text{rad/s}^2$

(b) $\dfrac{37.7 - 31.4}{1.91} = t = 3.30 \,\text{s}$

9.

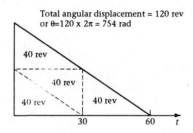

Total angular displacement = 120 rev
or θ=120 x 2π = 754 rad

40 rev

40 rev

40 rev    40 rev

30    60    $t$

(a) $\omega_0 = \left(\dfrac{2\theta}{t}\right) - \omega_1 = \left(\dfrac{2 \times 754}{60}\right) - 0 = 25.1 \,\text{rad/s} \ \text{ or } \ \dfrac{25.1}{2\pi} \times 60 = 240 \,\text{rev/min}$

(b) $\alpha = \dfrac{0 - 25.1}{60} = 0.418 \,\text{rad/s}^2$

10. (a) $\theta_A = \dfrac{\omega_1 \times 60}{2} = 30\omega_1, \ \theta_B = \dfrac{2\omega_1 \times 40}{2} = 40\omega_1$

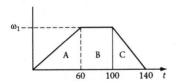

$\omega_1$

A    B    C

60   100   140   $t$

$\theta_C = \dfrac{\omega_1 \times 40}{2} = 20\omega_1, \quad \theta_{\text{total}} = \dfrac{\pi}{2} = \theta_A + \theta_B + \theta_C$

$\therefore \dfrac{\pi}{2} = 90\omega_1 \quad \therefore \omega_1 = 0.0175 \,\text{rad}$

(b) $\alpha = \dfrac{\omega_1}{t} = \dfrac{0.0175}{60} \doteq 0.000\,29 \,\text{rad/s}^2$

(c) $\dfrac{-\omega_0}{t} = \alpha = \dfrac{0.0175}{40} = 0.000\,44 \,\text{rad/s}^2$

11. (a) $\omega_0 = \dfrac{1500 \times 2\pi}{60} = 157 \,\text{rad/s}, \quad \omega_1 = \dfrac{750 \times 2\pi}{60} = 78.5 \,\text{rad/s},$

$t = \dfrac{157 - 78.5}{1.5} = 52.35 \quad \text{total time} = 52.33 + 30 + 20 = 102.33 \,\text{s}$

(b) $\theta_A = \dfrac{(157 + 78.5) \times 52.35}{2} = 6164.2 \text{ rad}, \quad \theta_B = \dfrac{(78.5 + 78.5)}{2} 30 = 2355 \text{ rad}$

$\theta_c = \dfrac{(157 + 78.5)20}{2} = 2355 \text{ rad}, \quad \text{Total } \theta = 6164.2 + 2355 + 2355$

$$= 10\,874 \text{ rad}$$

$$\text{or } \dfrac{10\,874}{2\pi} = 1730.7 \text{ rev}$$

12. (a) $\omega = \dfrac{5.0}{0.025} = 200 \text{ rad/s}$

(b) $\alpha = \dfrac{0.75}{0.025} = 30 \text{ rad/s}^2$

.

## CHAPTER   3

1. (a) $v_1 = \dfrac{40\,000}{3600} = 11.11 \text{ m/s}, \; a = \dfrac{-v_1}{t} = \dfrac{11.11}{15} = 0.74 \text{ m/s}^2$

$F = 3000 \times 0.74 = 2220 \text{ N}$

(b) $s = \dfrac{(11.11 + 0)15}{2} = 83.3 \text{ m}$

2. (a) acceleration upwards $= 9.81 + 2 = 11.81 \text{ m/s}^2, \; F = 90 \times 11.81 = 1063 \text{ N}$
   (b) acceleration downwards $= 9.81 - 2.5 = 7.31 \text{ m/s}^2, \; F = 90 \times 7.31 = 658 \text{ N}$

3. (a) $s = \dfrac{8.33^2}{2 \times 2.5} = 13.9 \text{ m},$

(b) $F = 22\,000 \times 2.5 = 55 \text{ kN}$

4. (a) $I = 400 \times 0.5^2 = 100 \text{ kg m}^2,$

(b) $\omega_1 = 6 \times 2\pi = 37.7 \text{ rad/s}, \; \alpha = \dfrac{-37.7}{40} = 0.94 \text{ rad/s}^2, \; T = 100 \times 0.94 = 94 \text{ N m}$

5. $\alpha = \dfrac{50.26 - 37.7}{40} = 0.32 \text{ rad/s}^2, \; T = 70 \times 0.4^2 \times 0.32 = 3.58 \text{ N m}$

6. (a) $I = (50 \times 0.75^2) + (5 \times 0.06^2) + (7 \times 0.6^2) + (900 \times 0.3^2) + (12 \times 0.75^2)$

$$= 118.4 \text{ kg m}^2$$

(b) $\alpha = \dfrac{942 - 0}{90} = 10.47 \text{ rad/s}^2, \; T = 118.4 \times 10.47 = 1240 \text{ N m}$

7. $(9000 \times 3.33) - (14\,000 \times 1.11) = (9000 + 14\,000)v_3$

$\therefore v_3 = 0.627 \text{ m/s}$

8. $v_3 = \dfrac{4.7 \times 10^6}{400} = 11\,750 \text{ m/s}, \quad v_2 = 11\,750 + 1000 = 12\,750 \text{ m/s}$

9. $(25 \times 0.393^2 \times 21) + 0 = [(25 \times 0.393^2) + (60 \times 0.406^2)]\omega_3$

$$\therefore \omega_3 = \frac{81}{13.75} = 5.9 \text{ rad/s or } 56 \text{ rev/min}$$

10. $(590.8 \times 15.7) + 648\omega_2 = (590.8 + 648) \times 6.28$

$$\omega_2 = \frac{-1495.9}{648} = -2.3 \text{ rad/s or } -22 \text{ rev/min}$$

## CHAPTER 4

1. Yes.

2.

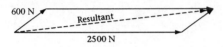

Resultant = 3050 N at 6° to horizontal

3.

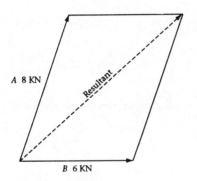

Resultant = 11.5 KN
at 41° to horizontal

4.

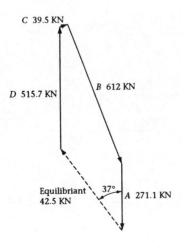

5.

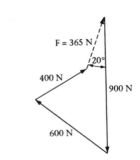

6.

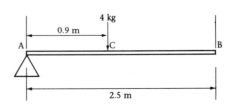

Moments about $A$:

$$B \times 2.5 = 4 \times 0.9 \times 9.81 = \frac{4 \times 0.9}{2.5} \times 9.81 = 14.13\,\text{N}$$

7.

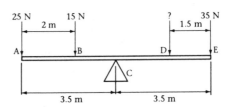

Moments about C:

$$(25 \times 3.5) + (15 \times 1.5) = (D \times 2) + (35 \times 3.5)$$

$$87.5 + 22.5 = 2D + 122.5$$

$$\frac{110 - 122.5}{2} = D = -6.25\,\text{N} \quad \text{(upwards)}$$

8.

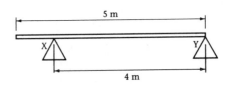

(a) $\dfrac{120}{5} = 24\,\text{kg/m}$

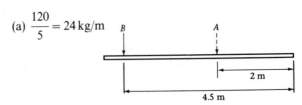

$$\therefore A = 4 \times 24 \times 9.81 = 942\,\text{N}$$

$$B = 1 \times 24 \times 9.81 = 235\,\text{N}$$

Moments about $Y$:

$$(235 \times 4.5) + (942 \times 2) = X \times 4$$

$$\frac{2942}{4} = X = 735\,\text{N}$$

$$\therefore \text{total load} = 1177\,\text{N} = X + Y$$
$$\text{Load at } Y = Y = 1177 - X$$
$$= 1177 - 735 = 442\,\text{N}$$

(b)

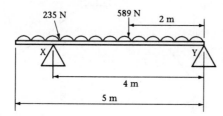

Since 589 N is equally divided between the two supports each support must be supporting 294.5 N:

$$\therefore X = 735 + 294.5 = 1029.5\,\text{N}$$

$$Y = 442 + 294.5 = 736.5\,\text{N}$$

9. (a)

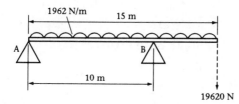

Moments about A:

$$B \times 10 = (15 \times 1962 \times 7.5)$$

$$B = \frac{220\,725}{10} = 22\,072.5\,\text{N}$$

$$A = 29\,430 - 22\,072.5 = 7357.5\,\text{N}$$

(b) Moments about A:

$$B \times 10 = (15 \times 1962 \times 7.5) + (19\,620 \times 15)$$

$$B = \frac{220\,725 + 294\,300}{10} = 51\,503\,\text{N}$$

Moments about B:

$$(A \times 10) + (19\,620 \times 5) + (1962 \times 5 \times 2.5) = 1962 \times 10 \times 5$$

$$A = \frac{-1962 \times 12.5}{10} = 2452.5\,\text{N} \quad \text{downwards}$$

10. (a) $\theta = 180° - 60° - 50° = 70°$

$A = 100\cos 70° = 34.2\,\text{N}$

$\therefore$ moment $= 34.2 \times 0.4 = 13.7\,\text{N m}$

(b) $60°$

## CHAPTER 5

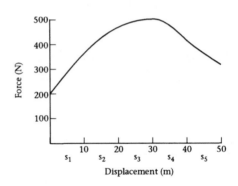

1. (a) $F_{av} = \dfrac{280 + 440 + 490 + 480 + 380}{5} = 414\,\text{N}$

(b) WD $= 415 \times 60 = 24.9\,\text{kJ}$

2. (a) WD $= 10 \times 7 = 70\,\text{kwh}$ or $70 \times 3.6 = 252\,\text{mJ}$

(b) $W = T\omega$ or $T = \dfrac{W}{\omega} = \dfrac{10 \times 10^3 \times 60}{600 \times 2\pi} = 159\,\text{N m}$

3. (a) $T = 250 \times 0.225 = 56.25\,\text{N m}$

(b) $W = T\omega = 56.25 \times \dfrac{680 \times 2\pi}{60} = 4\,\text{kW}$

(c) $\eta\% = \dfrac{4 \times 10^3}{4.3 \times 10^3} \times 100 = 93\%$

4. $a = \dfrac{v}{t} = \dfrac{15}{200} = 0.075\,\text{m/s}^2$, $s = \dfrac{(0 + 15)200}{2} = 1500\,\text{m}$

$W = \dfrac{\text{WD}}{t} = \dfrac{Fs}{t} = \dfrac{mas}{t} = \dfrac{500 \times 10^3 \times 0.075 \times 1500}{200} = 281.25\,\text{kW}$

5. $v = \dfrac{90 \times 1000}{3600} = 25\,\text{m/s}$, $v_V = 25\sin 8° = 3.48\,\text{m/s}$

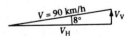

$W = Fv_V = 2000 \times 9.81 \times 3.48 = 68.26\,\text{kW}$

$\eta = \dfrac{\text{Power out}}{\text{Power in}}$ or $W_{in} = \dfrac{W_{out}}{0.7} = \dfrac{68.26}{0.7} = 97.5\,\text{kW}$

6. $WD = Fs = 3 \times 0.5 = 1.5\,\text{kJ}$

$$\text{Power} = \frac{WD}{t}$$

$$\therefore t = \frac{WD}{W} = \frac{1.5 \times 10^3}{80} = 18.75\,\text{s}$$

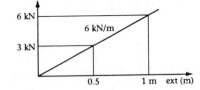

7. $W_{\text{out}} = F_0 v = 550 \times 9.81 \times 9 = 48.6\,\text{kW}$

$$\eta = \frac{48.6}{65} \times 100 = 75\%$$

## CHAPTER 6

1. $F = \mu N \quad \therefore \mu = \dfrac{F}{N} = \dfrac{90}{40 \times 9.81} = 0.23$

2. $F = \mu N = 0.25 \times 41 \times 9.81 = 100.5\,\text{N}$,

$$\mu = \frac{F}{N} = \frac{100.5}{70 \times 9.81} = 0.146$$

3. $P_{\text{HORIZ}} = P\cos 20° = 0.94P$, $P_{\text{VERT}} = P\sin 20° = 0.342P$,

$$N = (40 \times 9.81) - 0.342P = 392.4 - 0.342P$$

$$F = \mu N = 0.94P = 0.3(392.4 - 0.342P)$$

$$\therefore P = 113\,\text{N}$$

4. (a) $F = \mu N \therefore P\cos 20° = 0.4[(250 \times 9.81) - P\sin 20°] \therefore P = 911.3\,\text{N}$

   (b) $P\cos 10° = 0.4[(250 \times 9.81) + P\sin 10°] \therefore P = 1071.7\,\text{N}$

5. $\mu = \tan 25° = 0.466$, $P = 4905\sin 25° + 0.466 \times 4905\cos 25° = 4145\,\text{N}$

6. $50 = 500\sin 20° + \mu 500\cos 20° \therefore \mu = 0.26$

   $\text{up} = \text{down} \therefore P\cos 10° = (500\sin 20°) + \{0.26[(500\cos 20°) - (P\sin 10°)]\}$

   $$P = 284\,\text{N}$$

7. $\theta = \arctan\dfrac{10}{\pi \times 60} = 3°$, $\delta = \arctan 0.12 = 6.84°$

   $T = 6000 \times 0.03\tan(6.84 + 3) = 31.22\,\text{N m}$

8. (a) $\delta = \arctan 0.1 = 5.7°$, $\theta = \arctan\left(\dfrac{5}{\pi \times 60}\right) = 1.52°$

   $$\omega = \frac{2T}{D\tan(\theta + \delta)} = \frac{2 \times 6}{0.06\tan(1.52 + 5.7)} = 1579\,\text{N or } \frac{1579}{9.81} = 161\,\text{kg}$$

   (b) $T = 161 \times 0.03 \times \tan(5.7 - 1.52)$

## CHAPTER 7

1. (a) $f = \dfrac{1}{1.3} = 0.769\,\text{Hz}$,

   (b) $\omega = 2\pi \times 0.769 = 4.833\,\text{rad/s}$,

   (c) $\dot{x}_{max} = 4.833 \times 0.7 = 3.38\,\text{m/s}$

   $\ddot{x}_{max} = 4.833^2 \times 0.7 = 16.35\,\text{m/s}^2$

2. From $f = \dfrac{1}{2\pi}\sqrt{\dfrac{\ddot{x}}{x}}$, $x = \dfrac{\ddot{x}}{(2\pi f)^2} = \dfrac{8}{(2\pi \times 10)^2} = 2.02\,\text{mm}$

3. (a) $\ddot{x} = 12.56^2 \times 0.04 = 63.1\,\text{m/s}^2$

   (b) $0.15 = 0.4 \sin \omega t$

   $\sin \omega t = \dfrac{0.15}{0.4} = 0.375 \; \therefore \omega t = 0.384\,\text{rad}$

   $\dot{x} = \omega A \cos \omega t = 12.56 \times 0.4 \times \cos 0.384 = 4.65\,\text{m/s}$

4. $\ddot{x} = \dfrac{20}{20} = 1\,\text{m/s}^2,\; 1 = \omega^2 \times 0.02$

   $\therefore \omega^2 = 50$

   $\omega = 7.07$

   $f = \dfrac{7.07}{2\pi} = 1.13\,\text{Hz}$

5. $\omega = \dfrac{2\pi}{0.7} = 9\,\text{rad/s},\; \dot{x} = 9 \times 0.2 = 1.8\,\text{m/s},\; \text{KE} = \dfrac{2(1.8)^2}{2} = 3.24\,\text{J}$,

6. (a) $\omega = \dfrac{2\pi}{0.08} = 78.5\,\text{rad/s},\; t = \dfrac{\sin^{-1}(x/A)}{\omega} = \sin^{-1}\dfrac{(0.0035/0.006)}{78.5} = 7.93 \times 10^{-3}\,\text{s}$

   $\dot{x} = 78.5 \times 0.006 \cos(78.5 \times 7.93 \times 10^{-3}) = 0.38\,\text{m/s}$

   (b) $\ddot{x} = -78.5^2 \times 0.006 \sin(78.5 \times 7.93 \times 10^{-3}) = 21.56\,\text{m/s}^2$

7. (a) $\omega = \dfrac{3000 \times 2\pi}{60} = 31.4\,\text{rad/s},\; \tau = \dfrac{2\pi}{31.4} = 0.2\,\text{s},\; t = \dfrac{\tau}{4} = 0.05\,\text{s}$

   $\ddot{x} = -31.4^2 \times 0.02 = 19.7\,\text{m/s}^2,\; F = 4 \times 19.7 = 78.9\,\text{N}$

   (b) $\dot{x} = 31.4 \times 0.02 \times \cos(31.4 \times 8.1 \times 10^{-3}) = 0.6\,\text{m/s},\; \text{KE} = \dfrac{4 \times 0.6^2}{2} = 0.72\,\text{J}$

## CHAPTER 8

1. $\sigma = \dfrac{100 \times 10^3}{0.08 \times 0.03} = 41.7\,\text{MPa},\; \varepsilon = \dfrac{41.7 \times 10^6}{200 \times 10^9} = 2.08 \times 10^{-4}$

2. $\sigma = \dfrac{40 \times 10^3}{(\pi/4) \times 0.022^2} = 105.2\,\text{MPa}, \; x = \dfrac{2 \times 105.2 \times 10^6}{200 \times 10^9} = 0.001\,05\,\text{m} = 1.05\,\text{mm}$

3. 100 mm diameter:

$\sigma = \dfrac{500 \times 10^3}{(\pi/4) \times 0.1^2} = 63.7\,\text{MPa}, \; \varepsilon = \dfrac{63.7 \times 10^6}{200 \times 10^9} = 3.18 \times 10^{-4}$

60 mm diameter:

$\sigma = \dfrac{500 \times 10^3}{(\pi/4) \times 0.06^2} = 176.8\,\text{MPa}, \; \varepsilon = \dfrac{176.8 \times 10^6}{200 \times 10^9} = 8.84 \times 10^{-4}$

$\text{change in length} = 0.8 \times 3.18 \times 10^{-4} + 8.84 \times 10^{-4} = 11.38 \times 10^{-4}\,\text{m}$
$= 1.138\,\text{mm}$

4. $\sigma_c = \dfrac{500 \times 10^3 \times 0.025}{2 \times 0.001} = 6.25\,\text{MPa}$

5. $P = \dfrac{2 \times 0.003 \times 100 \times 10^6}{0.5} = 1.2\,\text{MPa}$

6. $\sigma = \dfrac{42 \times 10^3}{(\pi/4) \times 0.025^2} = 85.6\,\text{MPa}, \; \varepsilon = \dfrac{0.000\,63}{0.6} = 1.05 \times 10^{-3}$

$\varepsilon = \dfrac{85.6 \times 10^6}{1.05 \times 10^{-3}} = 81.5\,\text{GPa}$

$v = \dfrac{0.0067/25}{1.05 \times 10^{-3}} = 0.255$

7. $\sigma = \dfrac{75 \times 10^3}{0.05 \times 0.012} = 125\,\text{MPa}, \; \varepsilon = \dfrac{125 \times 10^6}{200 \times 10^9} = 6.25 \times 10^{-4}$

$\text{charge in length} = 6.25 \times 10^{-4} \times 800 = 0.5\,\text{mm}$
$\text{change in width} = -0.3 \times 6.25 \times 10^{-4} \times 50 = -0.0094\,\text{mm}$
$\text{change in thickness} = -0.3 \times 6.25 \times 10^{-4} \times 12 = -0.0023\,\text{mm}$

8. $\tau = \dfrac{1.6 \times 10^3}{2 \times (\pi/4) \times 0.006^2} = 28.3\,\text{MPa}$

9. $\sigma = \dfrac{10 \times 10^3}{(0.080 - 0.015) \times 0.012} = 12.8\,\text{MPa}$

$\tau = \dfrac{10 \times 10^3}{(\pi/4) \times 0.015^2} = 56.6\,\text{MPa}$

10. $F = 450 \times 10^6 \times \pi \times 0.04 \times 0.005 = 2.83 \times 10^5\,\text{N}$

$\sigma = \dfrac{2.83 \times 10^5}{(\pi/4) \times 0.04^2} = 225\,\text{MPa}$

## CHAPTER 9

1.

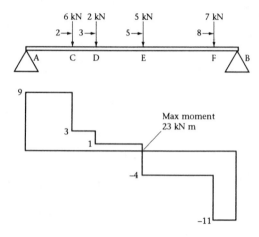

Reactions:

$$R_B \times 9 = (6 \times 2) + (2 \times 3) + (5 \times 5) + (7 \times 8):$$

$$R_B = 11\,kN$$

$$R_A = 6 + 2 + 5 + 7 - 11 = 9\,kN$$

$$M_C = 9 \times 2 = 18\,kN\,m$$

$$M_D = (9 \times 3) - (6 \times 1) = 21\,kN\,m$$

$$M_E = (9 \times 5) - (6 \times 3) - (2 \times 2) = 23\,kN\,m$$

$$M_F = (9 \times 8) - (6 \times 6) - (2 \times 5) - (5 \times 3) = 11\,kN\,m$$

2.

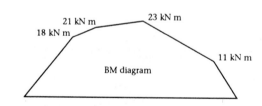

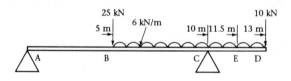

Reactions – moments about A:

$$C \times 10 = (25 \times 5) + (10 \times 13) + (8 \times 6 \times 4) = R_C = 44.7\,kN$$

$$R_A = 25 + 10 + (6 \times 8) - 44.7 = 38.3\,kN$$

$$M_B = (38.3 \times 5) = 191.5\,kN\,m$$

$$M_C = (38.3 \times 10) - (25 \times 5) - (6 \times 5 \times 2.5) = 183\,kN\,m$$

$$M_E = (38.3 \times 11.5) - (25 \times 6.5) - (6 \times 6.5 \times 3.25) = 151.2\,kN\,m$$

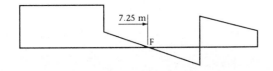

$$M_F = (38.3 \times 7.25) - (25 \times 2.25) - (6 \times 2.25 \times 1.13) = 206.17 \, \text{kN m}$$

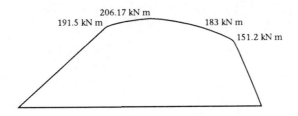

3.

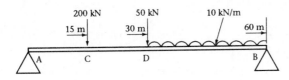

Reactions $60R_B = (200 \times 15) + (50 \times 30) + (10 \times 30 \times 45)$    $\therefore R_B = 300 \, \text{kN}$

$R_A = 200 + 50 + (10 \times 30) - 300 = 250 \, \text{kN}$

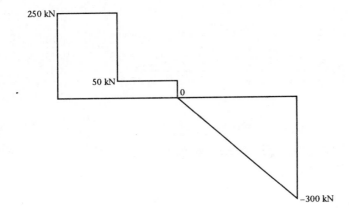

$$M_C = (250 \times 15) = 3750 \, \text{kN m}$$
$$M_D = (250 \times 30) - (200 \times 15) = 4500 \, \text{kN m}$$

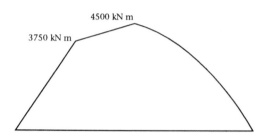

4.

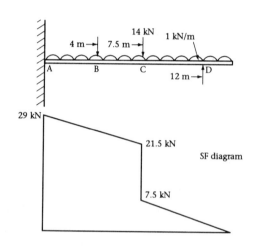

$$M_A = (14 \times 7.5) + (1 \times 15 \times 7.5) = 217.5 \, \text{kN m}$$
$$M_B = (14 \times 3.5) + (1 \times 11 \times 5.5) = 109.5 \, \text{kN m}$$
$$M_C = (1 \times 7.5 \times 3.75) = 28.125 \, \text{kN m}$$
$$M_D = (1 \times 3 \times 1.5) = 4.5 \, \text{kN m}$$
Max BM at built-in end

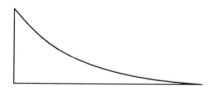

5.

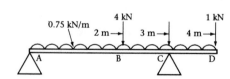

Reactions – moments about C:
$$(R_A \times 3) + (1 \times 1) + (0.75 \times 1 \times 0.5) = (4 \times 1) + (0.75 \times 3 \times 1.5)$$
$$3R_A + 1 + 0.375 = 4 + 3.375$$
$$R_A = \frac{6}{3} = 2\,kN$$

$$R_C = 4 + 1 + (0.75 \times 4) - 2 = 6\,kN$$

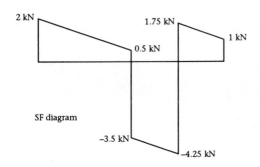

$$M_B = (2 \times 2) - (0.75 \times 2 \times 1) = 2.5\,kN\,m$$
$$M_C = (2 \times 3) - (0.75 \times 3 \times 1.5) - (4 \times 1) = -1.375\,kN\,m$$

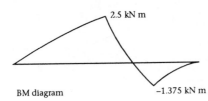

6.

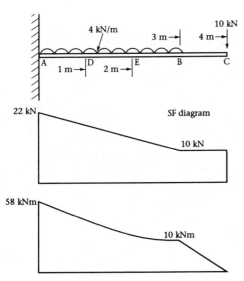

$$M_A = (4 \times 3 \times 1.5) + (10 \times 4) = 58\,kN\,m$$
$$M_D = (4 \times 2 \times 1) + (10 \times 3) = 38\,kN\,m$$

$$M_E = (4 \times 1 \times 0.5) + (10 \times 2) = 22 \, kN \, m$$
$$M_B = (10 \times 1) = 10 \, kN \, m$$
Max. SF $= 22 \, kN$
Max. BM $= 58 \, kN$

7.

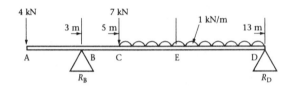

Reactions:
$$R_B \times 10 = (4 \times 13) + (7 \times 8) + (1 \times 8 \times 4) \therefore R_B = 14 \, kN;$$
$$R_D = 4 + 7 + (1 \times 8) - 14 = 5 \, kN$$

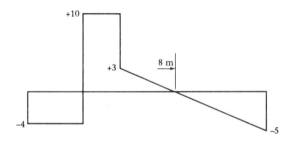

$$M_B = (4 \times 3) = 12 \, kN \, m$$
$$M_C = (4 \times 5) - (14 \times 2) = -8 \, kN \, m$$
$$M_E = (4 \times 8) - (14 \times 3) + (7 \times 3) + (3 \times 1 \times 1.5) = 15.5 \, kN \, m$$

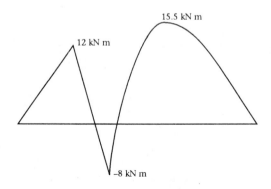

8.

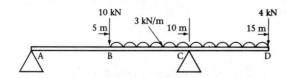

Reactions – moments about C:

$$R_A \times 10 = (10 \times 5) + (3 \times 5 \times 2.5) - (3 \times 5 \times 2.5) - (4 \times 5) \therefore R_A = 3\,\text{kN}$$
$$R_C = 10 + (3 \times 10) + 4 - 3 = 41\,\text{kN}$$

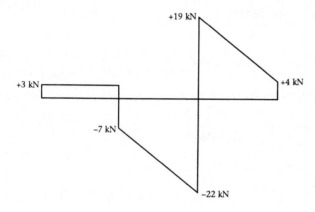

$$M_B = (3 \times 5) = 15\,\text{kN m}$$
$$M_C = (3 \times 10) - (10 \times 5) - (3 \times 5 \times 2.5) = -57.5\,\text{kN m}$$

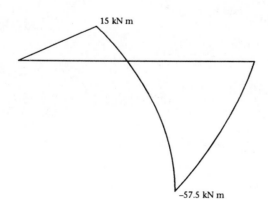

## CHAPTER 10

1. $M = 500 \times 2.5 = 1250 \, \text{N m}$

$$I = \frac{0.05 \times (0.12)^3}{12} = 7.2 \times 10^{-6} \, \text{m}^4$$

$$\sigma = \frac{1250 \times 0.06}{7.2 \times 10^{-6}} = 10.4 \, \text{MPa}$$

2. $\sigma = \dfrac{200 \times 10^9 \times 0.001}{0.6} = 333 \, \text{MPa}$

3. $I = \dfrac{\pi \times (0.001)^4}{4} = 0.785 \times 10^{-12} \, \text{m}^4$

$$M = \frac{333 \times 10^6 \times 0.785 \times 10^{-12}}{0.001} = 0.262 \, \text{N m}$$

4. $M = (2 \times 2) - (2 \times 1) = 2 \, \text{kN m}$

$$I = \frac{\pi}{4}(0.025^4 - 0.02^4) = 18.1 \times 10^{-8} \, \text{m}^4$$

$$\sigma = \frac{2000 \times 0.025}{18.1 \times 10^{-8}} = 276 \, \text{MPa}$$

5. $M = (10 \times 2.5) - (10 \times 1.25) = 12.5 \, \text{kN m}$

$$I = 2\left( \frac{0.1 \times (0.02)^3}{12} + (0.1 \times 0.02) \times 0.09^2 + \frac{0.01 \times (0.08)^3}{12} \right.$$

$$\left. + (0.01 \times 0.08) \times 0.04^2 \right) = 35.9 \times 10^{-6} \, \text{m}^4$$

$$\sigma = \frac{12\,500 \times 0.1}{35.9 \times 10^{-6}} = 34.8 \, \text{MPa}$$

6. $M = 3 \times 2 = 6 \, \text{kN m}$

$$I = \frac{6000 \times 0.1}{20 \times 10^6} = 30 \times 10^{-6} \, \text{m}^4$$

$$30 \times 10^{-6} = \frac{b \times (0.1)^3}{12}, \quad b = 0.36 \, \text{m}$$

7. (a) $J = \dfrac{\pi \times (0.025)^4}{2} = 0.614 \times 10^{-6} \, \text{m}^4$

$$\tau = \frac{500 \times 0.025}{0.614 \times 10^{-6}} = 20.4 \, \text{MPa}$$

(b) $\theta = \dfrac{500 \times 1}{0.614 \times 10^{-6} \times 80 \times 10^9} = 0.0102 \, \text{rad}$

8. $J = \dfrac{\pi \times (0.04)^4}{2} = 4.02 \times 10^{-6}\,\text{m}^4$

$\omega = \dfrac{2\pi \times 240}{60} = 25.1\,\text{rad/s}$

$T = \dfrac{25\,000}{25.1} = 2984\,\text{N m}$

$\tau = \dfrac{2984 \times 0.04}{4.02 \times 10^{-6}} = 29.7\,\text{MPa}$

9. $J = \dfrac{\pi}{2}(0.15^4 - 0.1^4) = 6.38 \times 10^{-4}\,\text{m}^4$

$T = \dfrac{40 \times 10^8 \times 6.38 \times 10^{-4}}{0.15} = 170\,170\,\text{N m}$

$W = 170\,170 \times \dfrac{120}{60} \times 2\pi = 2\,138\,414\,\text{W}$

$$= 2.14\,\text{MW}$$

10. $T = \dfrac{100 \times 10^3}{2\pi \times 30} = 530.5\,\text{N m}$

Shear stress:

$\dfrac{J}{r} = \dfrac{530.5}{50 \times 10^6} = 1.061 \times 10^{-5}\,\text{m}^3$

$\dfrac{J}{r} = \dfrac{\pi r^3}{2} \quad r = 0.0189\,\text{m}$

Angle of twist:

$\theta = \dfrac{2\pi}{360} = 0.0175\,\text{rad}$

$J = \dfrac{530.5 \times 1}{80 \times 10^9 \times 0.0175} = 0.379 \times 10^{-6}\,\text{m}^3$

$J = \dfrac{\pi r^4}{2} \quad r = 0.0222\,\text{m}$

Minimum diameter is 0.0444 m or 44.4 mm

## CHAPTER 11

1. $m = \dfrac{100 \times 10^3}{9.81} = 10\,193.7\,\text{kg}$

2. $P = 1030 \times 9.81 \times 70 = 707.3\,\text{kPa}$

3. $z = \dfrac{(101 - 99.4) \times 10^3}{13\,600 \times 9.81} = 0.012\,\text{m} = 12\,\text{mm}$

4. $z = 0.08 \times \sin 25° = 0.0338\,\text{m}$

$P = 1000 \times 9.81 \times 0.0338 = 331.7\,\text{Pa}$

5. $P = (13\,600 \times 9.81 \times 0.76) + (13\,600 \times 9.81 \times 0.22)$

$\quad = 130.74\,\text{kPa}$

6.

|   | $q$ | $w$ | $u_1$ | $u_2$ | $\Delta u$ |
|---|-----|-----|-------|-------|------------|
| a | 25 | 15 | 40 | 50 | 10 |
| b | 15 | $-10$ | 10 | 35 | 25 |
| c | $-20$ | $-10$ | 15 | 5 | $-10$ |
| d | 25 | 11 | 12 | 26 | 14 |

7. $w = 300 \times 10^3 (0.5 - 0.1) = 120\,\text{kJ}$

8. $w = 100 \times 10^3 \times \dfrac{4}{3}\pi(0.125)^3 = 818.1\,\text{J}$

9. $P + aV = b$ is a straight line curve.

$V_2 = \dfrac{(6 \times 10^5) - (500 \times 10^3)}{10^6} = 0.1\,\text{m}^3$

$w = \dfrac{(100 + 500)}{2} \times 10^3 (0.1 - 0.5) = 120\,\text{kJ}$

10. $(50 + q_2) = (35 - 40)$

$q_2 = -55\,\text{kJ}$

## CHAPTER 12

1. $R = 8.314/16 = 0.52\,\text{kJ/kg K}$

$m = \dfrac{500 \times 10^3 \times 0.5}{0.52 \times 10^3 \times 288} = 1.67\,\text{kg}$

2. $q = 0.75 \times 10^3 (353 - 293) = 45\,\text{kJ}$

$P_2 = 150 \times \dfrac{353}{293} = 180.7\,\text{kPa}$

3. $\rho = \dfrac{101 \times 10^3}{0.287 \times 10^3 \times 293} = 1.201\,\text{kg/m}^3$

$q = 1.201 \times (3 \times 4 \times 2) \times 720(20 - 10) = 207.5\,\text{kJ}$

$\tau = \dfrac{207.5}{2} = 103.75\,\text{s}$

4. $T_2 = 288\left(\dfrac{800}{100}\right)^{0.4/1.4} = 521.7\,\text{K}$

   $w = 720(521.7 - 288) = 168.3\,\text{kJ}$

5. $R = 8.314/28 = 0.297\,\text{kJ/kg K}$

   $\gamma = 1.04/(1.04 - 0.297) = 1.4$

   $P_2 = 1000\left(\dfrac{250}{500}\right)^{1.4/0.4} = 88.4\,\text{kPa}$

   $w = (1.04 - 0.297) \times 10^3(500 - 250) = 185.75\,\text{kJ}$

6. (a) $u = 762 + 0.7(2581 - 762) = 2035.3\,\text{kJ/kg}$

   (b) $x = \dfrac{2100 - 763}{2776 - 763} = 0.664$

7. $\Delta h = 2902 - (909 + 0.3(2797 - 909)) = 1426.6\,\text{kJ/kg}$

8. $q = (0.85 - 0.4)(2654 - 360) = 1032.3\,\text{kJ/kg}$

9. $\Delta u = (0.85 - 0.4)(2490 - 360) = 958.5\,\text{kJ/kg}$

   $w = q - \Delta u = 1032.3 - 958.5 = 73.8\,\text{kJ/kg}$

10. $\rho = \dfrac{1}{0.1} = 10\,\text{kg/m}^3$

    $x = \dfrac{1.129}{10} = 0.113$

    $u = 505 + 0.113(2529 - 505) = 733.7\,\text{kJ/kg}$

## CHAPTER 13

1. $1.2 \times \dfrac{\pi}{4}(0.1)^2 \times 20 = \rho_2 \times (0.05)^2 \times 12$

   $\rho_2 = 6.28\,\text{kg/m}^3$

2. $F = 1.2 \times \dfrac{\pi}{4}(2)^2 \times 60(70 - 50) = 4523.9\,\text{N}$

3. $v = \sqrt{(2 \times 25 \times 9.81)} = 22.1\,\text{m/s}$

   $F = 1000 \times \dfrac{\pi}{4}(0.05)^2 \times (22.1)^2 = 959\,\text{N}$

4. $q = (670 + 0.9(2756 - 670)) - 252 = 2295.4\,\text{kJ/kg}$

5. $T_2 = 293(8)^{0.4/1.4} = 530.8\,\text{K}$

   $w = 1005(293 - 530.8) = -238\,989\,\text{J/kg}$
   $= -239.0\,\text{kJ/kg}$

6. $v_2 = \sqrt{[2(2961 - (318 + 0.94(2637 - 318))) \times 10^3]} = 962.4\,\text{m/s}$

7. $Q = \left(1.2 \times \dfrac{\pi}{4}(0.05)^2 \times 15\right) \times 1005(60 - 20)$

   $= 1420.8\,\text{W}$

8. $v_2 = 25 \times (30/24)^2 = 39.1\,\text{m/s}$

   $P_2 = 100 \times 10^3 - \dfrac{1.2}{2}(39.1^2 - 25^2) = 99\,457.7\,\text{Pa}$

9. $\Delta P = 1000 \times 9.81 \times 0.014 = 137.3\,\text{Pa}$

   $v = \sqrt{\dfrac{2 \times 137.3}{1.25}} = 14.8\,\text{m/s}$

10. $\Delta P = 13\,600 \times 9.81 \times 0.015 = 2001.2\,\text{Pa}$

   $A_1/A_2 = 4$

   $v = 0.62\sqrt{\dfrac{2 \times 2001.2}{1000(4^2 - 1)}} = 0.32\,\text{m/s}$

   $\dot{m} = 1000 \times \dfrac{\pi}{4}(0.1)^2 \times 0.32 = 2.52\,\text{kg/s}$

## CHAPTER 14

1. $\tau = 0.06 \times \dfrac{20}{0.001} = 1200\,\text{Pa}$

   $F = 1200 \times 0.05 \times 0.2 = 12\,\text{N}$

2. $T = \dfrac{0.05 \times \pi^2 \times (0.03)^3 \times 0.015 \times 20}{2 \times 0.000\,05} = 0.04\,\text{N\,m}$

   $w = 0.04 \times 2\pi \times 20 = 5.02\,\text{W}$

3. $\text{Re} = \dfrac{1000 \times 1 \times 0.075}{0.001} = 75\,000$

4. (a) $\text{Re} = \dfrac{900 \times 4 \times 0.025}{0.1} = 900,\ \text{laminar}$

   (b) $\text{Re} = \dfrac{2.5 \times 4 \times 0.025}{0.000\,012} = 20\,833,\ \text{turbulent}$

   (c) $\text{Re} = \dfrac{1800 \times 4 \times 0.025}{0.05} = 3600,\ \text{transitional}$

5. $\Delta P = 2 \times 0.01 \times \dfrac{3000}{0.15} \times 1000 \times (0.5)^2 = 100 \times 10^3 \, \text{Pa}$

$$= 100 \, \text{kPa}$$

6. $\Delta P = \dfrac{32 \times 0.08 \times 10\,000 \times 0.4}{(0.4)^2} = 64\,000 \, \text{Pa}$

7. $\text{Re} = \dfrac{1000 \times 1 \times 0.015}{0.001} = 15\,000$

$f = 0.08 \times (15\,000)^{-1/4} = 0.0072$

$\Delta P = 2 \times 0.0072 \times \dfrac{100}{0.015} \times 1000 \times (1)^2 = 96\,384.2 \, \text{Pa}$

$\Delta z = \dfrac{96\,384.2}{1000 \times 9.81} = 9.83 \, \text{m}$

8. Small pipe:

$$f = 0.08 \times \left( \dfrac{1000 \times 1.2 \times 0.5}{0.001} \right)^{-1/4} = 0.002\,88$$

$\Delta P_{\text{S}} = 2 \times 0.002\,88 \times \dfrac{500}{0.5} \times 1000 \times (1.2)^2 = 8278.4 \, \text{Pa}$

Large pipe:

$v = 1.2 \times \dfrac{(0.5)^2}{(0.8)^2} = 0.469 \, \text{m/s}$

$$f = 0.08 \times \left( \dfrac{1000 \times 0.469 \times 0.8}{0.001} \right)^{-1/4} = 0.003\,23$$

$\Delta P_{\text{L}} = 2 \times 0.003\,23 \times \dfrac{700}{0.8} \times 1000 \times (0.469)^2 = 1244.3 \, \text{Pa}$

Total $\Delta P = 8278.4 + 1244.4 = 9522.8 \, \text{Pa}$

9. $\dot{m} = 1000 \times \dfrac{\pi}{4}(0.012)^2 \times 1.5 = 0.17 \, \text{kg/s}$

$$f = 0.08 \times \left( \dfrac{1000 \times 1.5 \times 0.012}{0.001} \right)^{-1/4} = 0.0069$$

$\Delta P = 2 \times 0.0069 \times \dfrac{35}{0.012} \times 1000 \times (1.5)^2 = 90\,650.7 \, \text{Pa}$

$\Delta P_0 = -(1000 \times 9.81 \times 10) - 90\,650.7 = -188\,750.7 \, \text{Pa}$

$W = \dfrac{0.17 \times (-188\,750.7)}{1000} = -32.1 \, \text{W}$

10. $\dot{m} = 1000 \times \dfrac{\pi}{4}(0.8)^2 \times 1.5 = 754 \, \text{kg/s}$

$$f = 0.08 \times \left( \dfrac{1000 \times 1.5 \times 0.8}{0.001} \right)^{-1/4} = 0.002\,42$$

$$\Delta P = 2 \times 0.002\,42 \times \frac{800}{0.8} \times 1000 \times (1.5)^2 = 10\,877\,\text{Pa}$$

$$\Delta P_0 = (1000 \times 9.81 \times 300) - 10\,877 = 2\,932\,110\,\text{Pa}$$

$$w = \frac{754}{1000} \times 2\,932\,123 = 22.1 \times 10^5\,\text{W}$$

$$= 2.21\,\text{MW}$$

## CHAPTER 15

1.  $A = 200/(15 \times (100 - 30)) = 0.19\,\text{m}^2$

2.  $A = 750/(30 \times (400 - 40)) = 0.0694\,\text{m}^2$
    $L = 0.0694/(\pi \times 0.0008) = 27.6\,\text{m}$

3.  $1500 = A((12 \times (150 - 25)) + 5.67 \times 10^{-8} \times (423^4 - 343^4))$

    $A = 0.593\,\text{m}^2$

4.  (a)  $Q/A = (0.8/0.128) \times (700 - 300) = 2500\,\text{W/m}^2$

    (b)  $h_c = 2500/(300 - 25) = 9.09\,\text{W/m}^2\,\text{K}$

5.  (a)  $\dfrac{1}{U} = \dfrac{0.128}{0.8} + \dfrac{0.025}{0.05} + \dfrac{0.012}{40} + \dfrac{1}{9.09}, U = 1.514\,\text{W/m}^2\,\text{K}$

    $Q/A = 1.514 \times (700 - 25) = 1022.3\,\text{W/m}^2$

    (b)  $1022.3 = 9.09 \times (T - 25), T = 137.5\,°\text{C}$

6.  $\dfrac{1}{U} = \dfrac{1}{10} + \dfrac{0.007}{1} + \dfrac{0.005}{0.026} + \dfrac{0.007}{1} + \dfrac{1}{10}, U = 2.46\,\text{W/m}^2\,\text{K}$

    $Q = 2.46 \times (0.25 \times 0.2) \times (200 - 25) = 21.5\,\text{W}$

7.  $\dfrac{1 \cdot}{U} = \dfrac{1}{3000} + \dfrac{0.002}{1} + \dfrac{1}{10}, U = 9.77\,\text{W/m}^2\,\text{K}$

    $A = (\pi \times 0.15 \times 0.2) + \left(2\dfrac{\pi}{4} \times 0.15^2\right) = 0.13\,\text{m}^2$

    $Q = 9.77 \times 0.13(100 - 20) = 101.3\,\text{W}$

8.  $\dfrac{1}{U} = \dfrac{1}{8} + \dfrac{0.008}{1} + \dfrac{0.015}{0.05} + \dfrac{0.003}{40} + \dfrac{1}{8}$

    $U = 1.79\,\text{W/m}^2\,\text{K}$

    $A = (4 \times 1 \times 0.5) + (2 \times 0.5 \times 0.5) = 2.5\,\text{m}^2$

    $Q = 1.79 \times 2.5(35 - (-5)) = 179\,\text{W}$

# References and suggested reading

Bedford, A. and Fowler, W. (1995) *Engineering Mechanics*, Addison-Wesley Publishing Co.

Beer, F.P. and Johnston, E.R. (1992) *Mechanics for Engineers*, McGraw-Hill.

Cain, J.A. and Hulse, R. (1990) *Structural Mechanics*, Macmillan.

Derry, T.K. and Williams, T.L. (1970) *A Short History of Technology*, Oxford University Press.

Hannah, J. and Hillier, M.J. (1988) *Applied Mechanics*, Longman.

Howatson, A.M., Lund, P.G. and Todd, J.D. (1991) *Engineering Tables and Data*, Chapman & Hall.

Rogers, C.F.C. and Mayhew, Y.R. (1988) *Thermodynamic and Transport Properties of Fluids*, Basil Blackwell.

Rolt, L.T.C. (1960) *George and Robert Stephenson*, Longman.

Rolt, L.T.C. (1967) *The Mechanicals*, Institution of Mechanical Engineers.

Sherwin, K. (1993) *Introduction to Thermodynamics*, Chapman & Hall.

Sherwin, K. and Horsley, M. (1995) *Thermofluids*, Chapman & Hall.

# Index